Lecture Notes in Artificial Intelligence   878

Subseries of Lecture Notes in Computer Science
Edited by J. G. Carbonell and J. Siekmann

Lecture Notes in Computer Science

Edited by G. Goos, J. Hartmanis and J. van Leeuwen

Toru Ishida

Parallel, Distributed and Multiagent Production Systems

Springer-Verlag

Berlin Heidelberg New York
London Paris Tokyo
Hong Kong Barcelona
Budapest

Series Editors

Jaime G. Carbonell
School of Computer Science, Carnegie Mellon University
Schenley Park, Pittsburgh, PA 15213-3890, USA

Jörg Siekmann
University of Saarland
German Research Center for Artificial Intelligence (DFKI)
Stuhlsatzenhausweg 3, D-66123 Saarbrücken, Germany

Author

Toru Ishida
Department of Information Science, Kyoto University
Kyoto 606-01, Japan

CR Subject Classification (1991): I.2.1, I.2.5, I.2.8, I.2.11, D.2.6, J.4, J.6-7

ISBN 3-540-58698-9 Springer-Verlag Berlin Heidelberg New York

CIP data applied for

Typesetting: Camera ready by author
SPIN: 10479235   45/3140-543210 - Printed on acid-free paper

Preface

This volume describes a series of research initiatives on concurrent production systems, which can be classified into three categories: (1) *Synchronous parallel production systems* or *parallel rule firing systems*, where rules are fired in parallel to reduce the total number of sequential production cycles, while rule firings are globally synchronized in each production cycle, (2) *asynchronous parallel production systems* or *distributed production systems*, where rules are distributed among multiple processes, and fired in parallel without global synchronization, and (3) *multiagent production systems*, where multiple production system programs compete or cooperate to solve a single problem or multiple problems.

Before going into the details of parallel, distributed and multiagent production systems, Chapter 1 first investigates the performance of a single production system. It has been known that as the scale of rule-based expert systems increases, the efficiency of production systems significantly decreases. To avoid this problem, recently developed production systems have been enabling users to specify an appropriate ordering or clustering of *join operations*, performed in the production system interpreters. Various *efficiency heuristics* have been introduced to optimize production rules manually. However, since the heuristics often conflict with each other, users have had to proceed by trial and error.

The problem addressed in Chapter 1 deals with how to estimate the efficiency of production system programs, and how to automatically determine efficient join structures. For this purpose, the *cost model for production systems* is introduced to estimate the run-time cost of join operations. An optimization algorithm is then provided based on the cost model. This algorithm does not directly apply efficiency heuristics to programs, but rather enumerates possible join structures under various constraints, and selects the best one. Evaluation results demonstrate that the optimization algorithm can generate programs that are as efficient as those obtained by manual optimization, and thus can reduce the burden of manual optimization.

For further speed-up of production system programs, parallel production systems are proposed in Chapter 2. The fundamental problem in synchronous parallel production systems is how to guarantee the serializability of production rule firings. *Interference analysis* is introduced to detect cases where a parallel firing result differs from the result of sequential firings of the same rules in any order. Based on a data dependency graph of production systems, general techniques applicable to both compile- and run-time analyses are provided.

Two algorithms are then proposed to realize the parallel rule firings on actual multiple processor systems. An efficient *selection algorithm* is provided to select multiple rules to be fired in parallel by combining the compile- and run-time interference analysis techniques. The *decomposition algorithm* partitions the given production system program and applies the partitions to parallel processes. A *parallel programming environment* is provided, including language facilities to allow programmers to make full use of the potential parallelism without considering the internal parallel mechanism. A parallel firing simulator is implemented to estimate how effective parallel firings of production system programs are. The effectiveness of parallel rule firings have been evaluated in several production system applications. Results show that the degree of concurrency can be increased by a factor of 2 to 9 by introducing parallel firing techniques. The sources of parallelism are investigated based on the evaluation results.

Distributed production systems are presented in Chapter 3. For asynchronous execution of production systems, parallel rule firing, where global control exists, is extended to distributed rule firing, where problems are solved by a society of production system agents using distributed control.

To perform domain problem solving in a distributed fashion, the agents need *organizational knowledge*, which represents both the necessary interactions among agents and their organization.

Organization self-design is then introduced into the distributed production systems to provide adaptive work allocation. Two *reorganization primitives, composition* and *decomposition*, are newly introduced. These primitives change the number of production systems and the distribution of rules in an organization. Organization self-design is useful when real-time constraints exist and production systems have to adapt to changing environmental requirements. When overloaded, individual agents decompose themselves to increase parallelism, and when the load lightens the agents combine with each other to free extra hardware resources. Simulation results demonstrate the effectiveness of this approach in adapting to changing environmental demands.

Chapter 4 investigates multiagent production systems, which are interacting multiple independent production systems, and thus are different from parallel or distributed production systems, which aim at improving the performance of a single production system program. A *transaction model* is introduced to enable multiple production systems to communicate through a shared working memory. To achieve arbitrary interleaved rule firings of multiple production system agents, each transaction is formed when a rule for firing is selected. An efficient concurrency control protocol, called the *lazy lock protocol*, is introduced to guarantee serializability of rule firings. As a result of allowing interleaved rule firings, however, ensuring the serializability becomes no longer enough to guarantee the consistency of the shared working memory information. A *logical dependency model* and its maintenance mechanisms are thus introduced to overcome this problem.

Finally, in Chapter 5, the control issues of multiagent production systems are further discussed. Because various events occur asynchronously in a multiagent network, the agents must cooperatively control their rule execution processes. A *meta-level control architecture* is required to prioritize the rule firings, to focus the attention of multiple agents on the most urgent tasks. To implement this, the procedural control of production rule firings is first introduced. The key idea of implementing the flexible control is to view production systems as a collection of independent rule processes, each of which monitors a working memory and performs actions when its conditions are satisfied by the working memory. *Procedural Control Macros (PCMs)*, which are based on Hoare's CSP, are then introduced to establish communication links between the meta-level control processes and the collection of rule processes. Although the PCMs are simple and easy to implement, through embedding them into procedural programming languages, users can efficiently specify the control plans for production systems.

The meta-level control architecture has been applied to construct a multiagent system called *CoCo*, which concurrently performs cooperative operations such as public switched telephone network control. The every day operations of telecommunication network operation centers are investigated, and then implemented in CoCo. The configuration of the CoCo agents, the plan description language, and experiments of multiagent rule firings are described.

This volume is written based on the following papers, which have already been published elsewhere. The author would like to thank the publishers and co-authors for permission to use these materials.

[Chapter 1: Production System Performance]

1. Toru Ishida, "Optimizing Rules in Production System Programs," *National Conference on Artificial Intelligence (AAAI-88)*, pp. 699-704, 1988.

2. Toru Ishida, "An Optimization Algorithm for Production Systems," *IEEE Transactions on Knowledge and Data Engineering*, Vol. 6, No. 4, pp. 549-558, 1994.

[Chapter 2: Parallel Production Systems]

3. Toru Ishida and Salvatore J. Stolfo, "Towards Parallel Execution of Rules in Production System Programs," *International Conference on Parallel Processing (ICPP-85)*, pp. 568-575, 1985.

4. Toru Ishida, "Methods and Effectiveness of Parallel Rule Firing," *IEEE Conference on Artificial Intelligence for Applications (CAIA-90)*, pp. 116-122, 1991.

5. Toru Ishida, "Parallel Firing of Production System Programs," *IEEE Transactions on Knowledge and Data Engineering*, Vol. 3, No.1, pp. 11-17, 1991.

[Chapter 3: Distributed Production Systems]

6. Toru Ishida, Makoto Yokoo and Les Gasser, "An Organizational Approach to Adaptive Production Systems," *National Conference on Artificial Intelligence (AAAI-90)*, pp. 52-58, 1990.

7. Les Gasser and Toru Ishida, "A Dynamic Organizational Architecture for Adaptive Problem Solving," *National Conference on Artificial Intelligence (AAAI-91)* , pp. 185-190, 1991.

8. Toru Ishida, Les Gasser and Makoto Yokoo, "Organization Self-Design of Distributed Production Systems," *IEEE Transactions on Knowledge and Data Engineering*, Vol. 4, No. 2, pp. 123-134, 1992.

[Chapter 4: Multiagent Production Systems]

9. Toru Ishida, "A Transaction Model for Multiagent Production Systems," *IEEE Conference on Artificial Intelligence for Applications (CAIA-92)*, pp. 288-294, 1992.

[Chapter 5: Meta-Level Control of Multiagent Systems]

10. Toru Ishida, "CoCo: A Multi-Agent System for Concurrent and Cooperative Operation Tasks," *International Workshop on Distributed Artificial Intelligence (DAIW-89)*, pp. 197-213, 1989.

11. Toru Ishida, Yutaka Sasaki and Yoshimi Fukuhara, "Use of Procedural Programming Languages for Controlling Production Systems," *IEEE Conference on Artificial Intelligence for Applications (CAIA-91)*, pp. 71-75, 1991.

12. Yutaka Sasaki, Keiko Nakata, Toru Ishida and Yoshimi Fukuhara, "Advantages of Meta-level Control Architectures in Maintaining Rule-Based Systems," *IEEE Conference on Tools with Artificial Intelligence (TAI-93)*, pp. 495-496, 1993.

13. Toru Ishida, Yutaka Sasaki, Keiko Nakata and Yoshimi Fukuhara, "An Meta-Level Control Architecture for Production Systems," *IEEE Transactions on Knowledge and Data Engineering*, 1995 (to appear).

This work would not have been possible without the contributions of a great many people. I wish to thank Seishi Nishikawa, Kunio Murakami, Tsukasa Kawaoka, Tetsuo Wasano, Ryohei Nakano, Fumio Hattori and Salvatore J. Stolfo for their support during this work at NTT Laboratories and Columbia University, and Les Gasser, Makoto Yokoo, Yutaka

Sasaki, Keiko Nakata and Yoshimi Fukuhara for their collaborative work. I also acknowledge Katsumi Tanaka, Shojiro Nishio, Kazuhiro Kuwabara, Yoshiyasu Nishibe, Daniel P. Miranker, James G. Schmolze and Hiroshi Okuno for their helpful discussions, Yuzou Ishikawa, Jun-ichi Akahani and Mark Lerner for providing the production system programs used in this study, and Satoshi Oyamada for introducing multiagent operations currently performed in telecommunication network operation centers.

September 1994   Toru Ishida

Contents

List of Figures

List of Tables

Chapter 1

Production System Performance

1.1 Introduction

The efficiency of production systems rapidly decreases as the number of working memory elements increases. This is because, in most implementations, the cost of join operations performed in the match process is directly proportional to the square of the number of working memory elements. Moreover, inappropriate ordering of conditions generates a large amount of intermediate data, which increases the cost of subsequent join operations.

To cope with the above problem, ART [Clayton, 1987], YES/OPS [Schor et al., 1987] and other production systems employ language facilities which enable users to specify an appropriate *join structure* (ordering or a clustering of join operations). However, since efficiency heuristics often conflict with each other, there is no guarantee that a particular heuristic always leads to optimization.

This chapter first describes production systems and their execution mechanism, called the RETE algorithm [Forgy, 1982]. The nature of production system performance is then examined: a *cost model* is introduced to estimate the run-time cost of arbitrary join structures. An optimization algorithm, which minimizes the total cost of join operations, is proposed based on the cost model. The algorithm does not directly apply the efficiency heuristics to the original rules, but rather enumerates possible join structures and selects the best one. Its basic methodology is to find effective constraints and to use those constraints to cut off an exponential order of possibilities.

The estimation is performed based on execution statistics measured from earlier runs of the program. The cost model is applied to possible join structures to estimate their run-time costs. All rules are optimized together so that join operations can be shared by multiple rules. Evaluation results demonstrate that the proposed algorithm can generate programs that are as efficient as those optimized by the expert system builder himself.

1.2   Production System

1.2.1   Architecture

This section provides a brief overview of production systems and an execution model of parallel rule firing.

A *production system* is defined by a set of *rules* or *productions*, called the *production memory (PM)*, together with a database of assertions, called the *working memory (WM)*. Assertions in the WM are called *working memory elements (WMEs)*. Each rule consists of a conjunction of *condition elements*, called the *left-hand side (LHS)* of the rule, along with a set of actions called the *right-hand side (RHS)*.

A rule written in the OPS5 production system language [Forgy, 1981] is shown below:

```
(p make-possible-trip
    (city ^name <x> ^state New-York)
  -(weather-forecast
        ^place <x> ^date tomorrow ^weather rainy)
    -->
    (make possible-trip ^place <x> ^date tomorrow)
```

The RHS specifies information which is to be added to or removed from the WM when the LHS is successfully matched with the contents of the WM. There are two kinds of condition elements: *positive condition elements* that are satisfied when there exists a matching WME, and *negative condition elements* that are satisfied when no matching WME is found. *Pattern variables* in the LHS are consistently bound throughout the positive condition elements. Thus the rule above may be read as:

```
If
    there is a WME in the system representing
    a city in New-York state
```

```
and
    there is no WME in the system representing that
    it will be rainy tomorrow in that city
then
    create a new WME tagging the city is a possible
    destination of tomorrow's trip.
```

Suppose the following is currently the only WME in the system:

```
(city ^name Buffalo ^state New-York)
```

where `city` represents a class name of WMEs, and attribute names are prefixed by '^' with attribute values immediately following. In this case, the LHS is satisfied, and if the rule is fired the following new WME is added to the system.

```
(possible-trip ^place Buffalo ^date tomorrow)
```

Throughout this book, *irrevocable forward chaining production systems* [Nilsson, 1980] including OPS5 will be focused attention on. The *production system interpreter* repeatedly executes the following cycle of operations:

1. *Match*:

   For each rule, determine whether the LHS matches the current environment of the WM.

2. *Select*:

   Choose exactly one of the matching instances of the rules, called *instantiations*, according to some predefined criterion, called a *conflict resolution strategy*.

3. *Act*:

   Add to or remove from the WM all assertions as specified by the RHS of the selected rule.

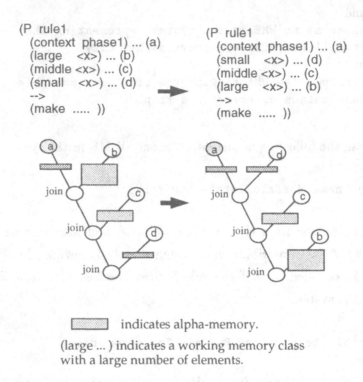

indicates alpha-memory.

(large ...) indicates a working memory class
with a large number of elements.

Figure 1.1: Heuristics 1: Place Restrictive Conditions First

1.2.2   The RETE Algorithm

In the RETE match algorithm [Forgy, 1982], the left-hand sides of rules
are transformed into a special kind of data-flow network called the *RETE
network*. The network consists of *one-input nodes, two-input nodes*, and ter-
minal nodes. The one-input node represents an *intra-condition test* or *selection*, which corresponds to an individual condition element. The two-input
node represents an *inter-condition test* or *join*, which tests for consistent
variable bindings between condition elements.

When a WME is added to (or removed from) the WM, a *token* which
represents the action is passed to the network. First, the intra-condition
tests are performed on one-input nodes; Assume the token is matched to
a positive condition element. The matched token is then stored in (or removed from) *alpha-memories*, and copies of the token are passed down to

```
(P  rule1                          (P  rule1
    (context  phase1) .... (a)          (context  phase1) .... (a)
    (frequently    <x>) .. (b)          (seldom        <x>) .. (d)
    (occasionally  <x>) .. (c)          (occasionally  <x>) .. (c)
    (seldom        <x>) .. (d)  ➡       (frequently    <x>) .. (b)
    -->                                 -->
    (make ..... ))                      (make ..... ))
```

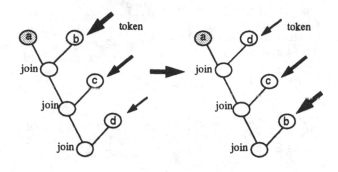

↙ indicates an amont of tokens.

(frequently ...) indicates a frequently changed
working memory class.

Figure 1.2: Heuristics 2: Place Volatile Conditions Last

successors of the one-input nodes. The inter-condition tests (*join* operations) are subsequently executed at two-input nodes. The tokens arriving
at a two-input node are compared against the tokens in the memory of the
opposite side branch. Then, paired tokens with consistent variable bindings
are stored in (or removed from) *beta-memories*, and copies of the paired
tokens are passed down to further successors. Tokens reaching the terminal
nodes are called *instantiations* and activate corresponding rules. A similar
process is performed when the token is matched to a negative condition
element. For more details, see [Forgy, 1982].

The following efficiency heuristics clearly identify the characteristics of
the RETE match algorithm [Clayton, 1987].

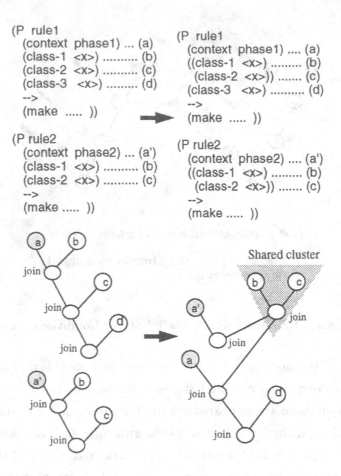

Figure 1.3: Heuristics 3: Share Join Clusters among Rules

Place restrictive conditions first.

It is known that the amount of intermediate data (*tokens*) dominates the cost of a join structure. One way to reduce the intermediate data is to join restrictive conditions first. Figure 1.1 illustrates the case where the conditions are reordered so that the WM class with the smaller number of WMEs comes first. This heuristic has been utilized to optimize conjunctive queries in AI and database systems [Smith and Genesereth, 1985; Warren, 1981].

Place volatile conditions last.

Suppose we have a rule with n conditions. If a newly created (or deleted) WME matches to the first condition, then $n - 1$ join operations would be invoked subsequently. However, if the matched condition is placed at the end of condition elements, only one join operation would be invoked. Thus, to reduce the number of join operations, volatile conditions should be placed last. Figure 1.2 represents the case where the conditions are reordered so that the frequently changed WM class comes last. This heuristic is peculiar to production systems. In database systems, since queries are optimized without considering the context, in which they are issued, such a dynamic feature has not been taken into account.

Share join clusters among rules.

If the same join cluster is shared by n rules, the cost of the join operations in the cluster would be reduced to $1/n$. To extract such sharable join clusters, rules are often required to transform their join structures. For example, in Figure 1.3, the second rule is transformed so that it can share $((b)(c))$ with the first rule. A similar idea has been discussed as *common subexpression isolation in multiple query optimization* [Jarke and Koch, 1984], but it is not yet common in database systems.

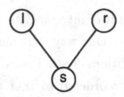

When r is not a negative condition element:

Test(s) = Token(l) x Memory(r) + Token(r) x Memory(l)
Token(s) = Test(s) x Ratio(s)
Memory(s) = Memory(l) x Memory(r) x Ratio(s)
Cost(s) = Cost(l) + Cost(r) + Test(s)

When r is a negative condition element:

Test(s) = Token(l) x Memory(r) + Token(r) x Memory(l)
Token(s) = Token(l) x Ratio(s)
Memory(s) = Memory(l) x Ratio(s)
Cost(s) = Cost(l) + Cost(r) + Test(s)

Figure 1.4: Cost Model

1.3   Cost Model

1.3.1   Parameters

The cost model of join operations is shown in Figure 1.4. Networks bounded by their lowest nodes are said to be *join structures of those particular nodes*. For example, the network shown in Figure 1.4 is a join structure of node s.

There are five parameters associated with each node n: $Token(n)$, $Memory(n)$, $Test(n)$, $Cost(n)$ and $Ratio(n)$. $Token(n)$ indicates the running total of tokens passed from node n to successor nodes. $Memory(n)$ indicates the average number of tokens stored in the alpha- or beta-memory of n. Note that $Token(n)$ and $Memory(n)$ are independent. $Token(n)$ represents the running total but $Memory(n)$ represents the average. Furthermore, $Token(n)$ increases monotonically, while $Memory(n)$ can decrease

WMEs are removed. $Test(n)$ indicates the running total of inter-condition tests at n. A consistency check of variable bindings between one arriving token and one token stored in a memory is counted as one test. $Cost(n)$ indicates the total cost of inter-condition tests performed in the join structure of n. The cost function is defined later. $Ratio(n)$ indicates the ratio of how often inter-condition tests are successful at n.

1.3.2 Equations

Let s be a two-input node joining two nodes, l and r, as shown in Figure 1.4. Note l and r are either one- or two-input nodes. The following equations are useful to determine the values of parameters.

1. $Test(s)$:

   When tokens are passed from the left, the number of tests performed at s is represented by $Token(l) \times Memory(r)$, and when from the right, $Token(r) \times Memory(l)$. Thus, $Test(s)$ is represented by $Token(l) \times Memory(r) + Token(r) \times Memory(l)$.

2. $Token(s)$:

   $Token(s)$ is represented by $Test(s) \times Ratio(s)$. However, when the right predecessor node is a negative one-input node, $Token(s)$ is represented by $Token(l) \times Ratio(s)$. This is because the negative condition element filters tokens passed from the left predecessor node.

3. $Memory(s)$:

   Analogous to database joins, $Memory(s)$ is represented by $Memory(l) \times Memory(r) \times Ratio(s)$. However, when the right predecessor node is a negative one-input node, $Memory(s)$ is represented by $Memory(l) \times Ratio(s)$.

4. $Cost(s)$:

   In general, the local cost at s can be represented by various functions. The cost functions should be adjusted to the production system interpreters. For example, for OPS5, since join operations are executed in a nested-loop structure, $Test(s)$, which represents the number of inter-condition tests, is appropriate. Thus, the cost of join structure s, $Cost(s)$, can be represented by $Cost(l) + Cost(r) + Test(s)$. For the system presented in [Gupta et al., 1987], on the other hand, $Token(s)$ might be better because hash tables are used.

1.3.3 Parameter Measurement

Production systems can measure some of the above parameters. When a production system program is executed, the values of $Test(n)$, $Token(n)$, and $Memory(n)$ for any given one or two-input node n can be recorded. Since $Test(n)$ and $Token(n)$ contain the running totals, the values are accumulated throughout execution. On the other hand, for $Memory(n)$, while the number of tokens in the alpha- or beta- memory is observed in each production cycle, the average number is calculated after execution. The production system interpreter can display the recorded parameters at any time upon request. Figure 1.5 represents an example of statistics for a sample production rule. This not only helps expert system builders to manually optimize the rule, but also provides the necessary initial values to the optimizer described in Section 1.4.

1.4   Performance Optimization

1.4.1   Motivation

Efficiency heuristics described in Section 1.1 cannot be applied independently, because the heuristics often conflict with one another. For example, applying one heuristic to speed up some particular rule can destroy shared join operations, slowing down the overall program [Clayton, 1987]. Thus, without an optimizer, expert system builders have to advance by a process of trial and error.

There are two more reasons why the development of a production system optimizer is necessary.

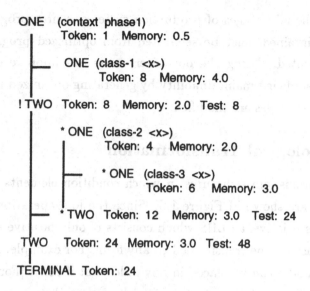

ONE (context phase1)
Token: 1 Memory: 0.5

ONE (class-1 <x>)
Token: 8 Memory: 4.0

! TWO Token: 8 Memory: 2.0 Test: 8

* ONE (class-2 <x>)
Token: 4 Memory: 2.0

* ONE (class-3 <x>)
Token: 6 Memory: 3.0

* TWO Token: 12 Memory: 3.0 Test: 24

TWO Token: 24 Memory: 3.0 Test: 48

TERMINAL Token: 24

ONE: one-input node
TWO: two-input node
! : inter-condition test with no join variables
* : shared node

Figure 1.5: Sample Statistics

1. *To enable expert system users to perform optimization:*

The optimal join structure depends on the execution statistics of production system programs. Thus, even if the rules are the same, results of the optimization may differ when different working memory elements are matched to the rules. For example, the optimal join structure for a circuit design expert system depends on the circuit to be designed. This means that the optimization task should be performed not only by expert system builders but also by expert system users. The optimizer can help users to tune expert systems to their particular applications.

2. *To improve efficiency without sacrificing maintainability:*

Production systems are widely used to represent expertise because of their maintainability. However, optimization sometimes makes rules unreadable by reordering conducted to reduce execution time. To preserve the advantages of production systems, source program files to be maintained must be separated from optimized program files to be executed. Using the optimizer, users can improve efficiency without sacrificing maintainability by generating optimized programs each time the rules are modified.

1.4.2   Topological Transformation

Examples of various join structures in which condition elements are variously clustered are shown in Figure 1.6. Since the join operation is commutative and associative, an LHS which consists of only positive condition elements can basically be transformed to any form. For example, in Figure 1.6, nodes *b*, *c* and *d* can be placed in any order, but if *MEA* [Forgy, 1981] is used as a conflict resolution strategy, node *a* cannot change its position. There are four topologies in Figure 1.6, and for each topology there exist $3! = 6$ structures. Therefore, in this case, 24 possible join structures, i.e., an exponential order of join structures, can exist.

When negative condition elements are present, there are a number of constraints in the transformation of a given LHS into an equivalent join structure. Since the role of negative condition elements is to filter tokens, they cannot be the first condition element, and all their pattern variables have to be bound by preceding positive condition elements. To simplify the following discussion, however, the detailed topological transformation constraints will not be explained. The optimization algorithm does not treat non-tree-type join topology, such as $(a)((b)(c))((b)(d))$, where two *b*'s share the same one-input node.

1.4.3   Algorithm Overview

An optimization algorithm presented here is not to directly apply the efficiency heuristics to the original rules, but rather to enumerate possible join structures and to select the best one. The basic methodology is to find effective constraints and to use those constraints to cut off an exponential order of possibilities. A *cost model* is utilized to estimate the run-time cost of join operations to be performed in possible join structures. The estima-

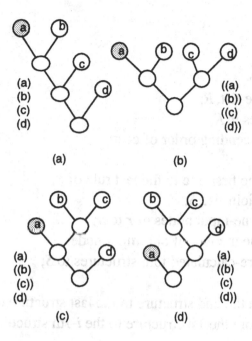

Figure 1.6: Topological Transformation

tion is performed based on execution statistics measured from earlier runs of the program. Furthermore, all rules are optimized together so that join operations can be shared by multiple rules.

Since there exists an exponential order of possible join structures, a simple generate-and-test method cannot handle this problem. The cost model of production systems permits the free combination of various topological transformations and the automatic estimation of their overall effects. The approach presented below is to generate join structures under various constraints, which reduce the possibilities dramatically. An outline of the algorithm is shown in Figure 1.7. The key points are as follows.

1. Sort rules according to their cost, measured from earlier runs of the program. Optimize rules one by one from higher-cost rules. This is done to allow higher-cost rules enough freedom to select join structures. (See multiple-rule optimization, described in Section 1.4.5.)

2. Before starting the optimization of each rule, the following nodes are

clear the rule-list, R;
push all rules to R;
sort R in descending order of cost;

for r from the first rule to the last rule of R;
    clear the join-list, S;
    push all one-input nodes of r to S;
    let k be the number of one-input nodes;
    append pre-calculated join structures to S;

    for i from the 2nd structure to the last structure of S;
        for j from the 1st structure to the i-1th structure of S;
            if
                all constraints are satisfied
            then do;
                create a join structure s to join i and j;
                calculate parameters of s;
                push s to just after the $max(i,k)$th structure of S
            end
        end
    end

    find the lowest-cost complete join structure;
    generate an optimized version of r
end

Figure 1.7: Optimization Algorithm

registered to the *join-list* of the rule: one-input nodes, each of which corresponds to a condition element of the rule, and two-input nodes in *pre-calculated join structures*, which are introduced to reduce search possibilities and to increase sharing join operations. The details of pre-calculated join structures are described in Section 1.4.5.

3. In the process of optimizing each rule, a two-input node is created by combining two nodes in the join-list. The created join structures are registered in the join-list if the same join structures have not already been registered. The algorithm chooses newer structures, which are possibly larger than the other registered structures, to accelerate the creation of a complete join structure of the rule. Constraints proposed later are used to reduce the number of possibilities.

4. After creating all possible join structures, the lowest-cost complete join structure is selected.

1.4.4 Parameter Estimation

In the process of optimization, various join structures are created and evaluated. The values of newly created two-input node parameters are calculated each time using the equations presented in Section 1.3.2.

To instantiate the equations, however, *Ratio(s)* for any given two-input node s should be determined. When the number of join variables is zero, the ratio is 1.0 because all Cartesian products are generated by such join operations. As the number of join variables increases, the ratio tends to decrease. However, since the value of *Ratio(s)* depends on the correlation between tokens to be joined, the accurate value of *Ratio(s)* is hard to obtain. Let *Variables(n)* be a set of pattern variables appearing in the join structure of n. When s is joining l and r, *Join_variables(s)* is defined by *Variables(l)* ∩ *Variables(r)*. To estimate *Ratio(s)*, it is assumed that the correlation among tokens depends on their join variables *Join_variables(s)*.

The procedure of calculating the ratio for any given two-input node is defined as follows:

1. *Measure the ratios for two-input nodes in given join structures.* Let $Ratio(s)$ be the observed ratio of two-input node s.

2. *Calculate the ratio for each pattern variable.* Let $Ratio(v_i)$ be the ratio of each pattern variable v_i. With the assumption of $Ratio(s) = \prod_i Ratio(v_i)$ where $v_i \in Join\_variables(s)$, calculate the ratios for each pattern variable, $Ratio(v_i)$, from the observed ratios, $Ratio(s)$, using the *least squared method.*

3. *Calculate the ratio for any two-input node created during the optimization process.* Let $Ratio(s')$ be the ratio of the created two-input node s' with the set of pattern variables $Join\_variables(s')$. Estimate $Ratio(s')$ by $\prod_j Ratio(v_j)$ where $v_j \in Join\_variables(s')$.

Figure 1.8 shows a trivial example to intuitively understand how the cost model yields the optimal join structure. In this example, Figure 1.8(a) represents an original join structure. Suppose one token successfully matches to condition element a, two to b, and one to c. Consequently, the same numbers of tokens are stored in alpha-memories. The observed ratios at two-input nodes are 1.0 for $((a)(b))$ and 0.5 for $(((a)(b))(c))$. The observed total cost is 4. By applying the above procedure, equations $Ratio(y) = 1.0$ and $Ratio(x) \times Ratio(z) = 0.5$ can be obtained, and thus $Ratio(x)$ and $Ratio(z)$ are estimated as 0.7. By using these ratios, the estimated cost of possible join structures are calculated as 2.4 for Figure 1.8(b) and 3.4 for Figure 1.8(c). Thus, Figure 1.8(b) is selected the optimal join structure according to the cost model.

To summarize, from the optimizer's point of view, $Token(a)$ and $Memory(a)$ are observed at any one-input node a and recorded for the subsequent optimization process. $Ratio(s)$ is also observed at any two-input node s, and utilized to calculate ratios of pattern variables. In the optimization process, $Ratio(s')$ for any created two-input node s' is estimated from the ratios of its pattern variables. Based on $Token(a)$ and $Memory(a)$ of any one-input node a and $Ratio(s)$ of any two-input node s, all other parameters in any created join structure are calculated using the equations presented in Section 1.3.2.

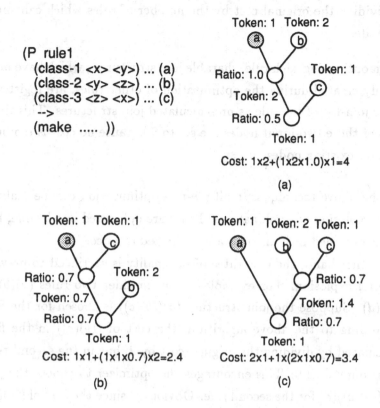

(a)

(b)   (c)

Figure 1.8: Optimal Join Structure

1.4.5 Multiple-Rule Optimization

Sharing join operations by multiple rules reduces the total cost of a program. However, examining all combinations of the sharing possibilities considerably increases the optimization time. Thus, the following techniques are used to increase the sharing opportunities, while retaining the simplicity of the optimization algorithm described in Section 1.4.3. In the following discussion, let $Conditions(n)$ be the set of condition elements included in the join structure of n, and $Variables(n)$ be the set of pattern variables appearing in $Conditions(n)$.

1. When creating a two input-node s, it is assumed that the join structure of s will be shared by all rules which contain $Conditions(s)$. The value of $Cost(s)$ is reduced based on this prediction: the cost is recalculated

by dividing the original cost by the number of rules which can share the node.

2. When optimizing each rule, sharable join structures , which have been already created during the optimization of other rules, are registered in the join-list of the rule as pre-calculated join structures. This time, costs of those two-input nodes are set to 0 because no cost is required to share existing nodes.

Using the above techniques, multiple-rule optimization can be realized without an explosion of combinations. Rules are optimized one by one, but the result is obtained as if all rules are optimized at once.

On the other hand, the guarantee of optimality is sacrificed to prevent combinatorial explosion. For example, let us consider two rules $(a)(b)(c)$ and $(b)(c)(d)$. Suppose the join structure $(a)((b)(c))$ is chosen for the first rule. According to the above algorithm, the cost of $((b)(c))$ in the first rule is calculated by dividing the original cost by 2, but in the second rule, the cost is counted as 0. This encourages the optimizer to choose the join structure $((b)(c))(d)$ for the second rule. Obviously, since the cost of $((b)(c))$ is significantly underestimated for the second rule, the algorithm cannot guarantee the optimality. From our experiences, however, since the rules are ordered from higher-cost ones, more freedoms are allowed in optimization of higher-cost rules, and thus better results tend to be obtained.

1.5   Constraints for Reducing Possibilities

The optimization algorithm described in Section 1.4, utilizes the following constraints to reduce the number of possible join structures.

1.5.1   Minimal-Cost Constraint

The *minimal-cost constraint* prevents the creation of a join structure whose cost is higher than that of the registered one. More formally, the constraint prevents the registration of the join structure of s to the join-list, if

<Production Rule>

```
(P  rule1
    (context  phase1) ..... (a)
    (class-1  <x>) ........... (b)
    (class-2  <x>) ........... (c)
    (class-3  <x>) ........... (d)
    -->
    (make ..... ))
```

<Join-List>

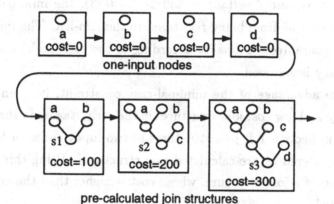

one-input nodes

pre-calculated join structures

<Minimal Cost Constraint>

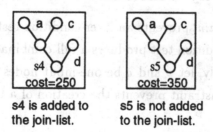

s4 is added to   s5 is not added
the join-list.   to the join-list.

Figure 1.9: Minimal-Cost Constraint

$\exists t \in$ *join-list*

such that $Conditions(s) \subseteq Conditions(t)$, and
$$Cost(s) \geq Cost(t).$$

Note that $Conditions(s) = \{s\}$, when s is a one-input node. In contrast, suppose the join structure of s is in the join-list, t is newly created, and s and t satisfy the above condition. Then, the join structure of s is replaced with that of t.

Figure 1.9 shows an example of the minimal-cost constraint. In this figure, the join-list already includes the join structure of $s3$, where $Conditions(s3)$ $= \{a, b, c, d\}$, and $Cost(s3) = 300$. Since $Conditions(s5) = \{a, c, d\} \subseteq Conditions(s3)$, and $Cost(s5) = 350 > Cost(s3)$, the minimal-cost constraints prevent $s5$ from being registered to the join-list. The minimal-cost constraint guarantees optimality according to the cost model if a tree-type join topology is assumed.

To take advantage of the minimal-cost constraint, it is important to create large and low-cost join structures in the early stages. In the algorithm described in Section 1.4, join structures of two-input nodes in the original rule are registered as pre-calculated join structures. Using this technique, the creation of a join structure, whose cost is higher than the original, can be prevented.

1.5.2 Connectivity Constraint

The *connectivity constraint* prevents an inter-condition test with no shared variable. Such inter-condition test produces a full combination of tokens to be joined. More formally, let p and q be one-input nodes in the currently optimized rule. The constraint prevents the creation of a two-input node s to join l and r, if

(i) $Variables(l) \cap Variables(r) = \emptyset$
    (i.e., $Join\_variables(s) = \emptyset$), and
(ii) $\exists p, q \notin Conditions(l) \cup Conditions(r)$
    such that $Variables(l) \cap Variables(p) \neq \emptyset$, and
            $Variables(r) \cap Variables(q) \neq \emptyset$.

```
<Production Rule>

    (P  rule1
        (context  phase1)   ...... (a)
        (class-1  <x>  <y>)  ...... (b)
        (class-2  <y>  <z>)  ...... (c)
        (class-3  <z>  <w>)  ...... (d)
        -->
        (make  ..... ))
```

<Connectivity Constraint>

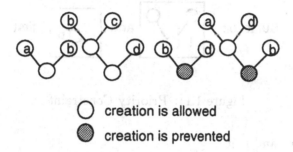

○ creation is allowed

◉ creation is prevented

Figure 1.10: Connectivity Constraint

In the example shown in Figure 1.10, the connectivity constraint prevents *b* and *d* from being joined because there is no shared variable (thus (i) is satisfied), and there remains a possibility of avoiding such a costly operation, if *b* and *c* or *c* and *d* are joined first (thus (ii) is satisfied). On the other hand, *a* and *b* are not prevented from joining, though there is no shared variable. This is because, sooner or later, *a* will be joined with some node without a shared variable anyway (thus (ii) is not satisfied).

1.5.3 Priority Constraint

It may be possible to prioritize the one-input nodes based on the execution statistics. The *priority constraint* prevents creating two-input nodes that join lower-priority nodes while higher-priority nodes can be joined. More formally, let p and q be one-input nodes, and $p \succ q$ indicates that p has a higher priority than q. The constraint prevents the creation of a two-input

<Priority>

Token(p) > Token(q)

and

Memory(p) > Memory(q)

<Priority Constraint>

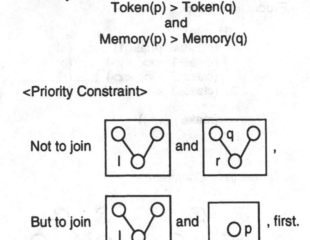

Not to join [ ] and [ ] ,

But to join [ ] and [ ] , first.

Figure 1.11: Priority Constraint

node s to join l and r, if

$$\exists p \notin Conditions(l) \cup Conditions(r),$$
$$\exists q \in Conditions(r)$$
such that (i) $p \succ q$, and
$$\text{(ii) } Variables(l) \cap Variables(p) \neq \emptyset, \text{ or}$$
$$Variables(l) \cap Variables(r) = \emptyset$$
$$\text{(i.e., } Join\_variables(s) = \emptyset).$$

At present, $p \succ q$ is defined only when $Token(p) > Token(q)$ and $Memory(p) > Memory(q)$. The equation (ii) is introduced to avoid the situation where joining l and p is prevented by the connectivity constraint while joining l and r is prevented by the priority constraint. Figure 1.11 illustrates the idea of the priority constraint. The major difference between the priority constraints and the *cheapest-first heuristics* [Smith and Genesereth, 1985] is that the priority constraint partially orders one-input nodes to exclude inappropriate join structures and thus can reduce the number of possible solutions, while the cheapest-first heuristics totally orders one-input nodes to directly produce a semi-optimal solution.

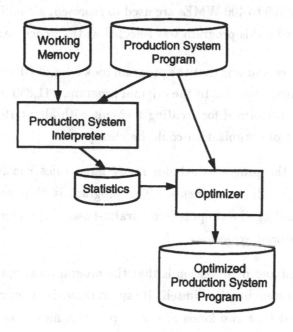

Figure 1.12: System Configuration

The connectivity and the priority constraints can significantly reduce the search possibilities, but sacrifice the guarantee of optimality.

1.6 Evaluation Results

1.6.1 Evaluation Environment

An optimizer applicable to OPS5-like production systems have been implemented. The system configuration is shown in Figure 1.12. A production system program to be optimized is executed once and its statistics are measured by a production system interpreter. The optimizer then inputs the program and its execution statistics, and outputs the optimized program. In this system, the overhead of statistics measurement is less than 5%, and thus the statistics are always recorded.

The optimizer is applied to a real-world production system program, a circuit design expert system [Ishikawa *et al.*, 1987]. This program consists of 107 rules, which generate and optimize digital circuits. In this evaluation,

approximately 300 to 400 WMEs are used to represent a circuit. There were
three reasons why this program was selected as the benchmark.

- The first reason was that the program took a lot of CPU time to create
  an optimized circuit. In the original program, 241,330 inter-condition
  tests are performed for creating a circuit with about 100 gates. Thus
  the effect of optimization could be clearly seen.

- Second, the program includes many large rules consisting of more
  than 20 condition elements. The program is thus not a mere *toy*
  for evaluating the proposed constraint-based approach to cope with
  combinatorial explosions.

- The third and main reason is that the program was optimized by the
  expert system builder himself. He spent three days optimizing it manually, and kept two kinds of source program files: one includes rules
  before optimization, and the other includes rules after optimization.
  Thus the effects of the proposed optimization algorithm can be compared to those of the manual optimization by applying the algorithm
  to the non-optimized rules.

1.6.2   Effects of Optimization

The result of optimizing the main module of the program, which consists
of 33 rules, is shown in Table 1.1. In this table, the cost of shared join
structures is calculated by dividing the measured cost by the number of
sharing rules. The results show that the total number of inter-condition
tests was reduced to 1/3, and CPU time to 1/2. Perhaps the most important
thing to note is that the optimizer produces a more efficient program than
the one obtained by manual optimization.

When carefully examining Table 1.1, readers may find that some inter-
condition tests of marked rules were increased in number by optimization.
These increases could be due to the inaccuracy of the cost model: though the
optimizer chooses the less costly join structures according to the cost model,
the actual cost can differ from the estimated one. However, the cases shown

Rule No.	Condition Elements	Before Optimization	After Optimization		Manual Optimization
		Inter-Condition Tests	Inter-Condition Tests	Created Join Structures	Inter-Condition Tests
1	21	46,432	28,824	280	47,888
2	18	29,548	505	170	27,245
3	17	29,548	505	72	27,245
4	22	25,513	144	185	7,813
5	21	25,313	144	93	7,813
6	18	10,322	3,749	443	3,749
7	17	10,322	3,749	184	3,749
*8	15	9,966	12,279	144	9,966
9	7	9,830	1,047	57	1,181
10	6	9,278	496	18	630
11	17	8,566	5,348	94	8,567
12	7	7,656	345	43	520
13	6	7,544	233	20	408
*14	23	3,160	7,152	99	4,616
15	11	1,918	2,165	71	1,918
16	20	1,336	1,337	236	1,336
17	19	845	846	117	845
*18	19	814	3,482	82	815
19	12	525	453	114	525
20	16	472	472	158	472
21	11	403	331	72	403
22	15	348	348	156	348
23	12	310	324	93	310
*24	16	253	253	118	253
25	11	202	216	63	202
26	20	181	181	146	181
27	19	181	181	97	181
*28	15	137	137	109	137
29	23	67	39	152	67
30	24	67	39	117	67
31	2	46	46	3	46
32	2	23	23	3	23
33	2	5	5	3	5
Total 33	Average 14.2	Total 241,330	Total 75,393	Average 115.5	Total 159,518
CPU Time (Ratio)		1.00	0.54		0.69

Table 1.1: Effects of Optimization

in Table 1.1 do not always mean that the optimization failed. For example, the optimizer creates *No.*1 and *No.*14 rules to share a large part of their join structures. Though the cost of *No.*14 rule increases after optimization, the total costs of the two rules has been decreased considerably.

Since optimization depends on the WM data, the obtained rules are not always efficient for any WM data. However, in the case of the circuit design expert system, users tend to run the expert system many times with almost the same WM data until a satisfactory circuit is obtained. The typical scenario is that the users run the expert system with the initial circuit data (probably including redundancies and bugs), optimize the rules suited to the circuit, and repeatedly run the optimized expert system to improve the circuit. Thus, even though optimization depends on the WM data, the optimizer is effective.

1.6.3   Effects of Constraints

The optimization time is directly proportional to the square of the number of created join structures. Thus the role of the constraints is to reduce the number of created join structures. The effectiveness of the constraints is shown in Table 1.2. Without the minimal-cost constraint, it is impossible to optimize rules which contain more than 10 condition elements. The connectivity and the priority constraints are also significant.

The number of join structures created during the optimization, and the number of run-time inter-condition tests of the resulting structures are also shown in Table 1.2. There are three cases depending on how many constraints are applied: the optimization is performed with all constraints (the minimal-cost, connectivity and priority constraints), with two constraints (the minimal-cost and connectivity constraints) and with only one constraint (the minimal-cost constraint). Note that, according to the cost model, the minimal-cost constraint preserves the optimal solution. Figure 1.13 illustrates the results listed in Table 1.1 and Table 1.2. The following observations are obtained from these results.

1. The number of inter-condition tests differs between the case where the connectivity/priority constraints are applied and the case where

Rule No.	Condition Elements	Minimal-Cost Connectivity Priority		Minimal-Cost Connectivity		Minimal-Cost	
		Inter-Condition Tests	Created Join Structures	Inter-Condition Tests	Created Join Structures	Inter-Condition Tests	Created Join Structures
1	21	28,824	280	28,824	>5,000	28,824	>5,000
2	18	505	170	493	850	494	2,451
3	17	505	72	493	93	494	129
4	22	144	185	136	1153	136	4,545
5	21	144	93	136	143	136	203
6	18	3,749	443	455	2,451	455	>5,000
7	17	3,749	184	455	109	455	>5,000
8	15	12,279	144	10,984	376	10,984	686
9	7	1,047	57	1,047	57	1,047	66
10	6	496	18	496	18	496	18
11	17	5,348	94	8,442	201	8,442	310
12	7	345	43	345	43	345	51
13	6	233	20	233	20	233	20
:	:	:	:	:	:	:	:

Table 1.2: Effects of Constraint

they are not: with application the number of run-time inter-condition tests increases by 10%. This shows that the connectivity and priority constraints should be applied only when a combinatorial explosion might occur during optimization.

2. On the other hand, the number of created join structures is 3.7 times higher without the priority constraint, and 6.3 times higher without the priority and connectivity constraints. Since the optimization time is proportional to the square of the number of the created join structures, the processing time is theoretically 14 to 40 times higher, respectively. However, the real overheads exceed the theoretical estimations due to such factors as garbage collection. As a result, the optimization of the circuit design program with only the minimal-cost constraint took 5 days on a Symbolics workstation. This shows that the constraints effectively reduce the number of possible solutions and thus shorten the optimization time.

The current strategy is to apply the priority constraints only to the

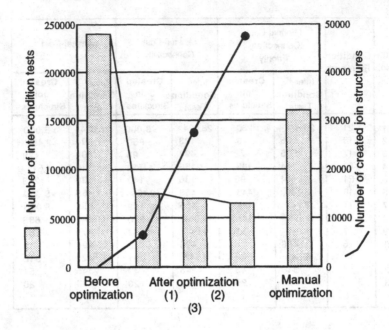

(1) with all constraints.
(2) with the Minimal-Cost and Connectivity constraints.
(3) with the Minimal-Cost constraint.

Figure 1.13: Summary of Evaluation

rules which have more than 10 condition elements. Under this strategy, the optimizer with all constraints takes somewhat more than 5 minutes to optimize the circuit design program on a Symbolics workstation. However, for average production system programs, in which the number of condition elements is only about 5 or so, optimization is usually completed in a few minutes.

1.7   Related Work

Compile-time optimization has been studied for conjunctive queries in AI and database areas [Smith and Genesereth, 1985; Warren, 1981; Jarke and Koch, 1984]. Various heuristics are investigated to determine the best ordering of a set of conjuncts. However, most of the previous studies on optimizing conjunctive queries are based only on statistics about sizes of the WM. Since production systems can be seen as programs on a database, statistics about run-time changes in the WM should also be considered. This makes optimization of production rules more complex than that of conjunctive queries.

To clarify the role of run-time WM changes in production system performance, the *cheapest-first heuristics* [Smith and Genesereth, 1985] has been applied to the circuit design expert system appeared in Section 1.6. This heuristics orders the condition elements based only on the sizes of the WM. The evaluation result shows that the *cheapest-first heuristics* miserably fails to create better join structures. The number of total inter-condition tests in the generated program is 333,397, which is 38% more than the original program, and CPU time increases 25%.

Another difference between optimizations of production systems and conjunctive queries can be found in the usage of heuristics. For example, the connectivity and priority constraints proposed in this chapter are respectively based on the *connectivity* and the *cheapest-first heuristics* described in [Smith and Genesereth, 1985]. However, the program behavior of production systems forces changes in the usage of those heuristics. In the query optimization, the heuristics are used to directly produce semi-optimal queries. In this chapter, however, the heuristics are modified to make them slightly weaker or less limiting, and use them as constraints to reduce the possibilities of join structures.

If the application requires only one execution to production system programs, however, the proposed algorithm cannot be applied. *Run-time optimization* techniques are suitable to such cases. The TREAT algorithm [Miranker, 1987] optimizes join operations dynamically. The method is called *seed-ordering*, where the changed alpha-memory is considered first, and the order of the remaining condition elements is retained. Since the overhead cannot be ignored for run-time optimization, sophisticated techniques such as those described here cannot be applied. The SOAR reorderer [Scales, 1986] attempts to directly apply the ordering heuristics for conjunctive queries to the run-time optimization of production rules. Unlike TREAT, since changes in the WM are not considered in the query optimization, applying the ordering heuristics to production rules often fails to produce better join structures. These results show that, even in runtime optimization, performance improvement cannot be achieved without considering WM changes.

1.8　Summary

The performance of production systems has been investigated based on the cost model, and an optimization algorithm has been proposed for production system programs. When applied to a circuit design expert system the algorithm produces a program that is as efficient as the one optimized by an expert system builder. As alternative approaches to increasing the performance of production systems, parallel and distributed production systems, which are described in Chapters 2 and 3, have been investigated. Since many of these studies have assumed the RETE pattern matching, the optimization algorithm proposed here is also effective in the parallel and distributed execution environments.

The cost model for production rules occupies a crucial part of this research, but as pointed out in Section 1.4.4, a completely accurate estimation is difficult to obtain. However, this does not mean we cannot do anything about the efficiency of production system programs. The proposed cost model has been successfully applied to a non-trivial example with more than 20 large rules. The results show that the optimized program becomes much faster than the original, and even faster than the manually optimized version. Though there may be cases where the proposed algorithm fails to improve efficiency, it is shown that the algorithm, on average, creates better join structures. The proposed algorithm will release both expert system builders and users from time-consuming optimization tasks.

Chapter 2

Parallel Production Systems

2.1 Introduction

Forward chaining production systems have been widely applied in the implementation of a number of knowledge-based problem solving systems. However, it has also been reported that the performance of production systems is not satisfactory when compared with the performance of systems with more conventional programming languages. Although advances in the implementation of production system interpreters have provided substantial performance improvements, further speed improvements are required for very large production systems with severe time constraints. To improve the efficiency of production systems, several multiple processor architectures have been investigated [Stolfo and Shaw, 1982; Forgy et al., 1984; Acharya and Tambe, 1989]. Two kinds of parallel algorithms were developed to more effectively utilize the parallel processing hardware.

One is *parallel rule matching* which aims to speed-up the matching process that consumes up to 90% of the total execution time. Decreasing the time to match rules makes it possible to compress each cycle in the execution of a production system. Gupta et al. [1986] have parallelized the *RETE match algorithm* [Forgy, 1982], and have reported that the average concurrency in actual expert systems has been improved 15.92-fold and the execution speed was increased 8.25-fold. Miranker [1987] has proposed the *TREAT match algorithm*, which was designed for fine-grain parallel processor systems.

Another type of parallel algorithm is *parallel rule firing*, which aims to reduce the total number of sequential production cycles by executing multiple matching rules simultaneously on a multiple processor system. The *SOAR production system language* [Scales, 1986] takes the parallel firing approach. In SOAR, each decision cycle consists of an elaboration phase and a decision procedure. The elaboration phase can fire rules in parallel, and creates objects and preferences. The decision procedure then fires rules sequentially examining the accumulated preferences and replacing objects. Gupta *et al.* [1988] have reported that SOAR's parallel firing mechanism increases the performance of production systems when combined with parallel rule matching.

This chapter applies the parallel firing approach [Ishida and Stolfo, 1985; Moldovan, 1986] to the *OPS5-like production system* [Forgy, 1981], which is the most widely used production system. The following three topics are mainly discussed to achieve the parallel firing approach.

Interference Analysis: In OPS5-like production systems, there are times when a parallel firing result differs from the result of any sequential firing. In those cases, we say there is *interference* among multiple rule firings. Production rules are originally written without consideration of interference with other rules. To guarantee the execution environment of a particular rule, it is necessary to determine what other rules need to be synchronized with the rule in question, and to suspend the firings of such rules during its execution. Therefore, general techniques are proposed based on a data dependency analysis for detecting interference both at compile-time and also at run-time.

Parallel Firing Algorithm: A *selection algorithm* that selects multiple rules to be fired in parallel is proposed. The compile- and run-time analysis techniques are combined to permit efficient and accurate interference detection. Since the interference can be analyzed roughly at compile-time, the run-time analysis process does not take much time. A *decomposition algorithm* is also proposed to find an optimal partition or distribution of given production rules onto multiple processor elements, so that the gain of parallel rule firing increases as much as possible.

Parallel Programming Environment: The difficulty in writing parallel production system programs stems from the lack of language facilities to accommodate parallel rule firings. Thus, the necessary facilities are included in the OPS5-like production system to enable users to make parallel programs without considering the internal parallel mechanism. A *parallel firing simulator* has been implemented to evaluate the potential parallelism of production system programs.

2.2 Parallel Firing Model

As described in Chapter 1, the production system interpreter repeatedly executes the *Match-Select-Act* cycle. In the parallel firing model, it is not assumed that only one rule is chosen in the Select phase. Rather, the firing of multiple rules simultaneously on multiple processors will be proposed.

When firing multiple production rules in parallel, however, there exists the case where the result of parallel execution of rules is different from the results of sequential executions in any order of applying those rules; the sequential firings are of the same rules but in random order. In this case, we say that there exists *interference* among multiple instantiation of rules. This concept is quite close to *the serializability of transactions* in the area of distributed databases [Bernstein and Goodman, 1981]. To avoid such an error, interference is detected in the Select phase.

Thus, in the parallel rule firing model, a production cycle is executed as follows.

Match: For each rule, determine whether the LHS matches the current environment of the WM.

Select: Choose as many instantiations as possible as long as interference does not occur among selected instantiations.

Act: Fire rules according to the selected instantiations simultaneously.

Since the parallel firing model allows multiple rules to be fired in parallel, programmers are thus not required to prioritize rules or assume any particular conflict resolution strategy. However, this restriction does not imply that we require *commutative production systems* [Nilsson, 1980], and is acceptable when combined with the parallel programming techniques described in Section 2.6.

2.3   Interference Analysis

2.3.1   Data Dependency Graph

To analyze the interference among multiple instantiations of production rules, a *data dependency graph of production systems* is introduced, which is constructed from the following primitives.

A *production node (a P-node)*, which represents a set of instantiations. P-nodes are shown as circles in all figures.

A *working memory node (a W-node)*, which represents a set of working memory elements. W-nodes are shown as squares in all figures.

A *directed edge from a P-node to a W-node*, which represents the fact that a P-node modifies a W-node. More precisely, the edge indicates that a WME in a W-node is modified (added or deleted) by the corresponding rule of an instantiation in a P-node. When a rule adds (deletes) WMEs to (from) a W-node, the W-node is said to be '+'*changed* ('-'*changed*) and the corresponding edge is labeled '+' ('-').

A *directed edge from a W-node to a P-node*, which represents the fact that a P-node refers to a W-node. More precisely, the edge indicates that a WME in a W-node is referenced by the corresponding rule of the instantiation in a P-node. When a WME is referenced by a positive (negative) condition element of a rule, the W-node is said to be '+'*referenced* ('-'*referenced*) and the corresponding edge is labeled '+' ('-').

```
(P make-possible-trip
    (candidate-city ^name <x> ^state New-York)
   - (weather-forecast
            ^place <x> ^date today ^weather rainy)
    -->
    (make  possible-trip ^place <x> ^date today)
    (remove 1))


(P make-weather-forecast
    (symptom ^animal frog ^action croak ^place <y>)
    (candidate-city ^name <y>)
    -->
    (make  weather-forecast
            ^place <y> ^date tomorrow ^weather rainy))
```

Figure 2.1: Sample Production Rules

The interference analysis, presented in this section, is effective for any size of P- or W-node. However, the larger the size is, the less information can be obtained. Since parallel rule firings are performed in a conservative manner, it would be better to associate P- and W-nodes with sets that are as small as possible.

2.3.2   Compile-Time and Run-Time Analyses

The interference among rule firings is analyzed by using a data dependency graph. The compile-time and run-time analyses are described below, using the rules in Figure 2.1.

Compile-Time Analysis

Figure 2.2 displays the data dependency graph constructed at compile-time. A P-node is associated with a set of instantiations of each rule. In Figure 2.2(a), a W-node is associated with a set of WMEs in each class. In Figure 2.2(b), by contrast, a W-node is associated with a set of WMEs represented by a matching pattern, which appears in the source production

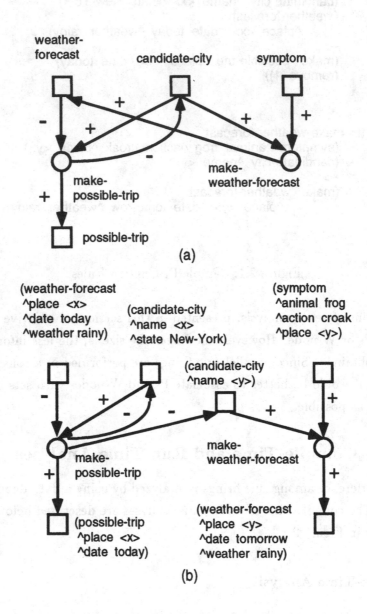

Figure 2.2: Compile-Time Analysis

system programs. The latter is a more general approach than the former: it can accommodate cases in which class names are represented by variables. Furthermore, by utilizing all information written in the source programs, more inherent parallelism in the program is extracted.

However, in the latter approach, a method to analyze overlapping W-nodes is necessary. Edges should be drawn from any P-node to all W-nodes that are overlapped by an RHS pattern associated with the P-node. For example, in Figure 2.2(b), since W-nodes, (`candidate-city ^name <x> ^state New-York`) and (`candidate-city ^name <y>`), overlap, '-'labeled edges are drawn from the P-node, `make-possible-trip`, to both W-nodes.

Run-Time Analysis

The run-time analysis can produce more accurate data dependency graphs than those of the compile-time analysis. Suppose there exist the following three WMEs.

```
(candidate-city ^name Buffalo   ^state New-York)
(candidate-city ^name New-York ^state New-York)
(symptom ^animal frog ^action croak ^place Buffalo)
```

A data dependency graph of the three created instantiations is shown in Figure 2.3. In this case, a P-node represents one instantiation, and a W-node represents an instantiated matching pattern. Since all pattern variables have already been instantiated, each W-node basically indicates a unique WME.

2.3.3   Paired-Rule and All-Rule Conditions

Paired-Rule Conditions

The following observations can be derived from a data dependency graph.

- If all W-nodes lying between `ruleA` and `ruleB` are '+'changed ('-'changed) by `ruleA` and '+'referenced ('-'referenced) by `ruleB`, then the firing probability of `ruleB` increases monotonically by executing `ruleA`. Thus, even if `ruleA` is fired during the execution of `ruleB`, interference never occurs.

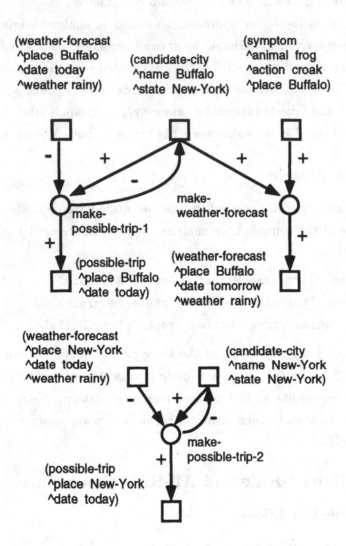

Figure 2.3: Run-Time Analysis

- Conversely, if some W-nodes lying between `ruleA` and `ruleB` are '+'changed ('-'changed) by `ruleA` and '-'referenced ('+'referenced) by `ruleB`, then the execution environment of `ruleB` may be destroyed by the firing of `ruleA`.

  For example, in Figure 2.4(a), suppose there initially exist two WMEs, `(class1)` and `(class2)`. If rules are fired sequentially, there remains `(class1)` when `ruleA` is first executed, or `(class2)` when `ruleB` is first executed. However, if two rules are fired in parallel, there remains no WME.

- If `ruleA` and `ruleB` change the same W-node, and if the W-node is '+'changed ('-'changed) by `ruleA` and '-'changed ('+'changed) by `ruleB`, then the result of the simultaneous firing may be different from the result of any sequential executions of `ruleA` and `ruleB`.

  For example, in Figure 2.4(b), suppose there is no WME before execution. If the rules are fired sequentially, there remains no WME when `ruleA` is first executed, or there remains two WMEs, `(class1)` and `(class2)`, when `ruleB` is first executed. If the rules are fired in parallel, however, there are four possibilities, i.e., `(class1)` and `(class2)` remain, `(class1)` remains, `(class2)` remains, or no WME remains.

From the above observations, the *paired-rule (or pairwise) conditions* [Ishida and Stolfo, 1985], which can detect the interference between instantiations in paired P-nodes, are formally described as follows.

[Paired-Rule Conditions]

There is a possibility of interference between the two instantiations in P-node P_A and P-node P_B, if there exists a W-node that satisfies any of the following conditions:

A1) '+'changed ('-'changed) by P_A and '-'referenced ('+'referenced) by P_B.

A2) '+'changed ('-'changed) by P_B and '-'referenced ('+'referenced) by P_A.

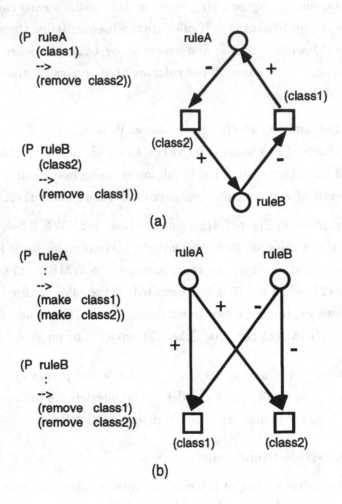

```
(P  ruleA
    (class1)
    -->
    (remove class2))
```

```
(P  ruleB
    (class2)
    -->
    (remove class1))
```

(a)

```
(P  ruleA
    :
    -->
    (make  class1)
    (make  class2))
```

```
(P  ruleB
    :
    -->
    (remove  class1)
    (remove  class2))
```

(b)

Figure 2.4: Paired-Rule Conditions

A3) '+'changed ('-'changed) by P_A and '-'changed ('+'changed) by P_B.

In OPS5, '+/-'changed WMEs is required to be '+'referenced in the LHS. Thus, in OPS5, condition *A3* is not necessary. However, a more general case is considered in the above conditions, such that WMEs can be added or deleted in the RHS without being referenced in the LHS.

All-Rule Conditions

If all P-nodes could be analyzed at once, however, more accurate results can be obtained by using the *all-rule (or cyclic) conditions*. An example of how this might be done is shown in Figure 2.5(a). Here, all rules can be safely fired in parallel even though the possibility of interference is detected by the paired-rule conditions. For instance, condition *A1* is satisfied between `ruleA` and `ruleB`, and between `ruleB` and `ruleC`. However, if there is no other rule, `ruleA`, `ruleB`, and `ruleC` can be fired in parallel, because in this case the result of parallel firing is the same as the sequential firing in the order `ruleC` $\Longrightarrow$ `ruleB` $\Longrightarrow$ `ruleA`. On the other hand, Figure 2.5(b) shows an example where parallel rule firing should not be employed. In this case, since `ruleA`, `ruleB`, and `ruleC` interfere with each other in a cyclic fashion, the result of parallel firing could differ from the results of any sequential firing. More formally, the all-rule conditions are described as follows.

[All-Rule Conditions]

Let $P_1...P_iP_{i+1}...P_{n+1}$, where $P_1 = P_{n+1}$, be a cyclic sequence of an arbitrary number of P-nodes. There is *cyclic interference* in P-nodes, if, for all i, there exists a W-node that is '+'changed ('-'changed) by Pi and '-'referenced ('+'referenced) by P_{i+1}. Interference occurs between two instantiations in P-nodes P_A and P_B, if any of the following conditions is satisfied:

B1) There exists cyclic interference in P-nodes that include P_A and P_B.

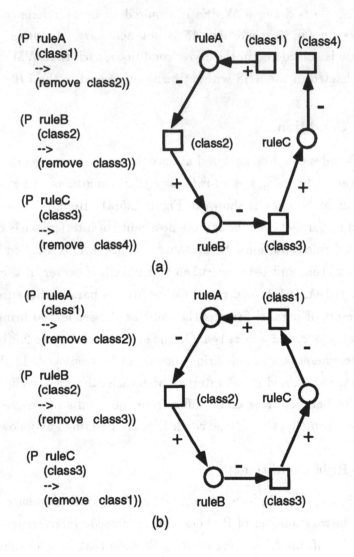

Figure 2.5: All-Rule Conditions

Conditions / Data dependency graph	Paired-rule conditions	All-rule conditions
Compile-time analysis (1) (Fig. 2.2(a))	(a) interference (b) interference (c) safe	(a) interference (b) interference (c) safe
Compile-time analysis (2) (Fig. 2.2 (b))	(a) interference (b) interference (c) safe	(a) safe (b) interference (c) safe
Run-time analysis (Fig. 2.3)	(a) some are safe (b) safe (c) safe	(a) safe (b) safe (c) safe

(a) Interference between an instantiation of 'make-possible-trip' and an instantiation of 'make-weather-forecast'.
(b) Interference between instantiations of 'make-possible-trip'.
(c) Interference between instantiations of 'make-weather-forecast'.

Table 2.1: Interference Analysis Techniques

B2) There exists a W-node, which is '+'changed ('-' changed) by P_A and '-'changed ('+'changed) by P_B.

In OPS5, since '+/-'changed WMEs is required to be '+'referenced in the LHS, condition $B2$ is not necessary.

2.4 Selection Algorithm

2.4.1 Accuracy of the Interference Analysis

Table 2.1 summarizes the results of various interference analyses on the rules shown in Figure 2.1. The variations are derived both from the preciseness of data dependency graphs (compile-time / run-time) and from the scope of interference detection (paired-rule / all-rule). The following points can be drawn from Table 2.1.

- Run-time analysis can permit more parallelism than compile-time analysis due to the preciseness of the P- and W-nodes. For example, run-time analysis concludes that the instantiations of `make-possible-trip` can be fired in parallel.

- The all-rule conditions can permit more parallelism than the paired-rule conditions. For example, the all-rule conditions conclude that the instantiations of `make-possible-trip` and `make- weather-forecast` can be fired in parallel.

Oshisanwo and Dasiewicz [1987] have also pointed out that not all the effects of parallelism can be determined by compile time techniques: excessive interference is indicated by compile-time techniques, which are inherently conservative. On the other hand, overheads cannot be ignored in any run-time analysis. In this section, compile- and run-time analysis techniques are combined to get a reasonably accurate and efficient selection algorithm.

2.4.2   Algorithm Overview

From the above discussions, run-time analysis using the all-rule conditions seems to be the best solution. However, the computational cost of checking the all-rule conditions is quite high, because condition *B1* requires that all strongly connected regions be detected from a data dependency graph. Since this overhead cannot be ignored in the run-time analysis, the efficient selection algorithm shown in Figure 2.6 is proposed by simplifying the all-rule conditions. For OPS5, the conditions in parentheses are not necessary. The key ideas are as follows:

- Generally, it is too expensive to build a data dependency graph at run-time. For practical purposes, we can say that there is no cyclic interference if, for all i, j where $i < j$, the i-th instantiation does not interfere with the j-th instantiation. Thus, the approach that does not make a complete data dependency graph at run-time, but checks instantiations one by one is selected. Note that the selection algorithm does not investigate the reverse direction: it is not necessary to check if the j-th instantiation interferes with the i-th instantiation.

- To reduce the overhead of the run-time analysis, the execution of the *Match-Select-Act* phases of the production cycle should be overlapped.

let *P1...Pn* be P-nodes each of which indicate
an instantiation created in the Match phase;
let *S* be a set of P-nodes to be fired in parallel;
S <- ∅;

do *i* = 1 to *n*;
   if
       the compile-time analysis guarantees
       Pi and P-nodes in *S* do not interfare with each other
   then
       add *Pi* to *S*
   else
      if
         Pi does not '+'refer to (or '+'change) a W-node
         which is '-'changed by P-nodes in *S*
         and
         Pi does not '-'refer to (or '-'change) a W-node
         which is '+'changed by P-nodes in *S*
      then
         add *Pi* to *S*
      end
   end
end

Figure 2.6: Selection Algorithm

Since the selection algorithm investigates instantiations one by one, it is possible to fire each selected instantiation immediately.

The proposed selection algorithm exposes less parallelism but is much more economical than examining the all-rule conditions. Compared with the paired-rule conditions, in which all interference between each instantiation pair are examined, the selection algorithm can permit more parallelism and is less expensive, because it does not require a check of the reverse direction.

It is also a good idea to incorporate compile-time analysis into the selection algorithm as shown in Figure 2.6. There are two advantages. First, the compile-time analysis can help the selection algorithm reduce run-time efforts: if two rules were determined not to cause interference at compile-time, a precise check can be avoided at run-time. Second, the compile-time analysis can help the selection algorithm perform a more accurate interference detection: in the compile-time analysis, though a data dependency graph is comparatively rough, the all-rule conditions can be completely examined.

2.5   Decomposition Algorithm

2.5.1   Mapping Rules on Multiple Processors

If a dynamic process allocation facility is not available, for example as in most distributed memory parallel machines, production systems have to be decomposed and allocated onto multiple processor elements *(PEs)* before execution. The aim of this decomposition is to map rules to a set of distinct PEs in such a way that the gain of parallel rule firing increases as much as possible.

The decomposition problem has three fundamental difficulties. First, the number of possible decompositions is too large to permit an exhaustive approach. Second, the optimal decomposition may change, when the usage of the production system program changes. Third, the gain of parallel rule firing depends on many factors: communication costs among PEs, reduction of production cycles, distribution of matching processes, etc. To overcome these difficulties, the following strategies are utilized.

- The *gain of parallel rule firing* is defined between each pair of rules so that it can be easily calculated by allocating the two rules on distinct PEs.

- To reduce the complexity of the combinatorial problem, the decomposition is performed based on the gain of each pair of rules. To obtain an efficient solution in an incremental manner, the most influential rule pair, which has the largest gain, is first allocated.

- Sample execution traces are used to calculate the gain of each pair of rules. Thus, analyzing additional execution traces incrementally approaches an optimal decomposition. Furthermore, tuning is possible after the production system starts to work.

2.5.2 Algorithm Overview

The *hierarchical decomposition algorithm* consists of two phases. In the first phase, the algorithm produces a hierarchical data structure, called a *decomposition tree*. In the second phase, partitions for parallel processor systems are created from the decomposition tree. Precise procedures in each phase are described below.

Phase 1: Generating a Decomposition Tree

A *token* is defined as a triple, *(ruleA ruleB G(ruleA,ruleB))*, where rule A and rule B are production rules and *G(ruleA,ruleB)* is the gain of parallel rule firing between rule A and rule B. It is assumed that all tokens for all pairs of rules are already calculated by using sample execution traces.

The goal of this phase is to produce a decomposition tree in which each rule is associated with a distinct leaf node, and to maximize the sum of $G(i,j)$ at each non-leaf node in all combinations of i and j, where i/j indicates the rule to be the right/left subtree of the non-leaf node. The tree initially consists of only one root node, which contains all of the rules. Tokens are input into the root node in the descending order of G. Each node processes tokens as follows.

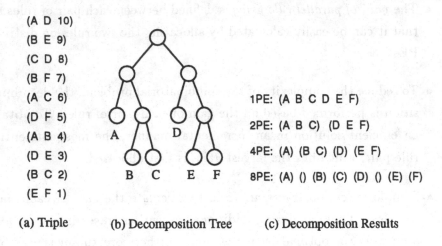

<table>
(A D 10)
(B E 9)
(C D 8)
(B F 7)
(A C 6)
(D F 5)
(A B 4)
(D E 3)
(B C 2)
(E F 1)
</table>

(A D 10)
(B E 9)
(C D 8)
(B F 7)
(A C 6)
(D F 5)
(A B 4)
(D E 3)
(B C 2)
(E F 1)

1PE: (A B C D E F)

2PE: (A B C) (D E F)

4PE: (A) (B C) (D) (E F)

8PE: (A) () (B) (C) (D) () (E) (F)

(a) Triple   (b) Decomposition Tree   (c) Decomposition Results

Figure 2.7: Decomposition Process

C1) *If both rules are contained in the root node,* move one of them to the root node of its left subtree, and the other to the root node of its right subtree. (If subtrees are null, create root nodes for those subtrees, and do the above operation.)

C2) *If one of two rules is contained in the root node and the other is contained in its right (left) subtree,* move the rule contained in the node to the root node of its left (right) subtree. (If the subtree is null, create the root node for the subtree, and do the above operation.)

C3) *If both rules are each contained in its left (right) subtree,* pass the token to the root node of its left (right) subtree.

C4) *If one of two rules is contained in its right subtree and the other is contained in its left subtree,* ignore this token.

Phase 2: Create Partitions for Multiple PEs

This phase creates partitions assigning rules to PEs from the decomposition tree. Because the rule pairs with large gain are decomposed at an early stage, partitions for a parallel processor system can be easily obtained by selecting a suitable level of the decomposition tree. The decomposition tree grows exponentially and thus is best mapped onto processors containing a binary power number of PEs. An example of the decomposition process is shown in Figure 2.7.

2.5.3 Decomposition Heuristics

In the decomposition algorithm presented above, tokens are processed in descending order. However, more heuristics for ordering tokens can be applied, because partitions derived in descending order have the following disadvantages.

- Sometimes, there exists a big difference between the number of rules on the right subtree and that on the left subtree. (For example, see Figure 2.8(a).)

- Rule pairs are often allocated without considering the relationship with other rule pairs. As a result, condition $C3$ ('*if both rules are contained in the node*') is often satisfied. (For example, see Figure 2.8(b).)

A more sophisticated ordering, called *the linked order of tokens*, has been invented to avoid the above disadvantages. In the linked order, tokens are ordered so that rules are allocated to right and left subtrees one after the other. The linked order is obtained by the following procedures.

1. Make a list of tokens in the descending order.

2. Pop a token from the list and register it as the first token of a linked order. Post both rules in the first token to a common table.

3. The n-th token of a linked order is selected as follows.

   (a) Check tokens in the list from the top to the end, and select the first token that contains a posted rule.

      i. If there is a token that contains some posted rule, then remove the token from the list and register it as the n-th token of the linked order. Next, find a rule in the n-th token that

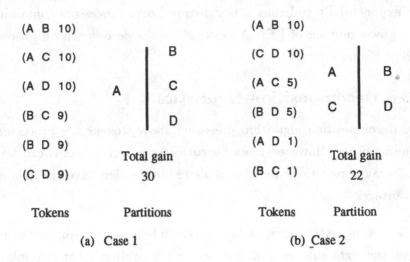

Figure 2.8: Decomposition with the Descending Order Heuristics

      is not posted. Clear the common table and post the found rule.

  ii. If there is no token, which contains any posted rule, then pop one token from the list and register it as the n-th token. Clear the common table and post both rules in the n-th token.

Figure 2.9 illustrates the decomposition results obtained by using the linked order for the same problems appearing in Figure 2.8. The sums of gain in both cases are improved by introducing the linked order. The linked order is applied to all levels of a decomposition tree by stacking and link-ordering all tokens, which are passed from a parent node. For the production system programs listed in Table 2.2, the linked-order decomposition produced partitions that were about 20% faster than the ones obtained by the decomposition based on descending order.

The number of possible pairs of rules is $n(n-1)/2$, when the number of rules is n. However, in an actual environment, the number of tokens is much smaller. The reason is that many pairs of rules will, in practice, never fire in parallel. For the production system programs in Table 2.2, only 17% of all possible pairs were processed as tokens.

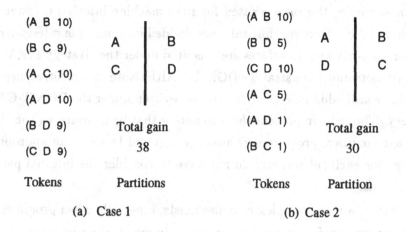

Figure 2.9: Decomposition with the Linked Order Heuristics

2.6 Parallel Programming Environment

2.6.1 Language Facilities

To fire multiple rules in parallel, it is essential that the rules be written without assuming a particular conflict resolution strategy. However, a conflict resolution strategy sometimes enables programmers to simplify rules. For example, by assuming the *MEA* or *LEX* strategy [Forgy, 1981], programmers can simplify the LHS of the rule, which is to be fired only when other rules cannot be fired. It is thus impractical to disable conflict resolution strategies. Rather, an alternative approach is taken by introducing the following language facilities:

- A *ruleset* is introduced to form a group of rules. Distinct conflict resolution strategies can be defined independently for each ruleset.

- A new conflict resolution strategy, *DON'T-CARE*, is introduced along with MEA and LEX. Rules in a ruleset under the strategy DON'T-CARE are fired in parallel.

For example, the rules written for man-machine interfaces, cannot be fired in parallel. To cover this, rules are divided into multiple rulesets: rulesets for man-machine interfaces are executed under the strategy MEA, and other rulesets under the strategy DON'T-CARE. Note that positive/negative conditions and add/ delete actions can be written under the DON'T-CARE strategy. The most important thing to note is that by introducing the above language facilities, programmers are only required to select an appropriate strategy for each ruleset, and do not have to consider the internal parallel mechanisms.

However, a more complex case also exists, especially when programs are originally written for sequential execution, in which the two types of rules are mixed: control rules shift the execution stages and heuristic rules solve the problem. Consider an instance where the control rules are to be executed under the MEA strategy, and the heuristic rules can be executed under the DON'T-CARE strategy. The following mechanism has been devised to accommodate cases of this kind.

- A *focusing mechanism* is introduced to transfer control from one ruleset to another. For example, suppose the rules are divided into two rulesets: CONTROL, for control rules, and HEURISTICS for heuristic rules. Then, the *focus* function enables the production system interpreter to perform the mixed execution of both rulesets.

        (focus HEURISTICS CONTROL)

Since higher priority is given to the former ruleset, the control rules to shift stages are fired only when the heuristic rules can no longer be fired.

2.6.2 System Configuration

To evaluate the effectiveness of parallel firing of production systems, a simulation environment has been developed. The environment consists of an *Analyzer*, a *Decomposer* and a *Simulator*.

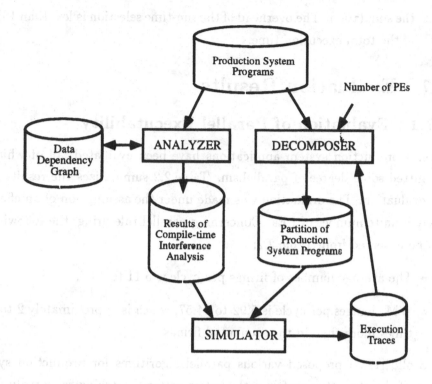

Figure 2.10: Parallel Programming Environment

The Analyzer inputs rules written in an OPS5-like production system. It constructs a data dependency graph of given production rules, analyzes the graph and outputs the results, which show what rules can be fired in parallel. The compile-time interference analysis, which is described in Section 2.3, is performed in the Analyzer.

The Decomposer analyzes the execution traces, which are created by the Simulator, and measures the gain of parallel rule firing between each pair of rules. The Decomposer creates partitions of production rules based on the hierarchical decomposition algorithm described in Section 2.5.

The Simulator simulates the parallel firing of production systems using the Selection algorithm described in Section 2.4. The results of the compile-time analysis performed by the Analyzer are referenced during

the simulation. The overhead of the run-time selection is less than 10%
of the total execution time.

2.7 Evaluation Results

2.7.1 Evaluation of Parallel Executability

Several production system applications have been evaluated, all of which
permitted some degree of parallelism. Table 2.2 summarizes the results of
the evaluation. The evaluation was made under the assumption of an effectively infinite number of PEs. Concerning parallel rule firing, the following
can be observed from Table 2.2:

- The average number of firings per cycle is 5.11 to 7.57.

- WM changes per cycle is 5.92 to 24.57, which is approximately 2 to 9
  times more than in the sequential firings.

Stolfo [1984] proposed various parallel algorithms for production systems. If the algorithm performs the entire pattern matching repeatedly for
each production cycle, it is expected that the speed to be improved by a
factor of 5.11 to 7.57. On the other hand, if the RETE match algorithm
is parallelized, the degree of concurrency can be increased by a factor of 2
to 9, because the number of WM changes (thus tokens) to be processed in
parallel is increased.

This number shows only the effect of parallel firing, i.e., it does not
include the effects of parallel matching, which may compress the production
cycle itself. Thus more speed enhancement than shown in Table 2.2 can be
expected from the total effect of parallel execution.

2.7.2 Sources of Parallel Executability

Parallelism in production system programs heavily depends on the nature
of the problem addressed. Some production system programs, such as the
Monkey and Bananas program, do not permit parallel firings at all. The
potential parallelism in production system programs can be classified as
follows:

Items / Programs	Sequential Firings			Parallel Firings					
				Sequential Ruleset		Parallel Ruleset			
	Rules	Cycles	WM Changes / Cycle	Rules	Cycles	Rules	Cycles	Firings / Cycle	WM Changes / Cycle
Circuit Design [Ishikawa et al., 1987]	107	518	2.75	56	324	51	38	5.11	22.32
Manhattan Mapper [Lerner and Cheng, 1983]	98	371	2.70	47	159	51	28	7.57	24.57
Waltz [Winston, 1977]	31	207	1.25	6	130	25	13	5.92	5.92
SPILL [Hayes-Roth et al., 1983]	24	58	4.27	3	11	21	9	5.22	10.44

Table 2.2: Evaluation of Parallel Executability

Rule parallelism: Multiple rules are fired in parallel without communication. For example, in the *Waltz labeling program* [Winston, 1977], which is included in Appendix B, many constraints can be applied independently. In the *circuit design program* [Ishikawa *et al.*, 1987], which has been developed at NTT laboratories, many rules for optimizing a circuit can be applied in parallel. The rule parallelism increases as the number of independent rules becomes larger.

Pipeline parallelism: Multiple rules are fired in parallel, passing data in a pipeline fashion. For example, in *Manhattan Mapper* [Lerner and Cheng, 1983], which has been developed at Columbia University to provide travel schedules in Manhattan, the length of the pipe is six: some rules are fired to generate candidate paths, and then other rules evaluate them. In the Waltz labeling program, though there is no clear pipe, the data modified by some rules are further modified by other rules. The pipeline parallelism increases as the length of the pipe becomes longer.

Data parallelism: Multiple instantiations of the same rule are fired in parallel based on distinct data. For example, in the Waltz labeling program, multiple instantiations of the same constraint rule are fired to label multiple edges in parallel. In the circuit design program, multiple instantiations of the same optimization rule are fired in parallel to optimize various portions of a circuit. The data parallelism increases as the number of WMEs becomes larger.

In the case of Manhattan Mapper, the parallel executability was improved many times in the evaluation process by avoiding the interference based upon results of the interference analysis. The initial evaluation of the original Manhattan Mapper show only a 15% reduction in production cycles, and that up to 2 rules are fired in parallel. This result was contrary to our expectations. After further investigation of the original Manhattan Mapper, it turned out that the system had been made suitable for sequential execution on single processor systems, i.e., *the execution order of rules was carefully pre-specified by embedded control information.*

To reveal the potential parallelism in Manhattan Mapper, the system was then reconstructed by analyzing its data dependency graph and by minimizing interference among rules. New results show more than an 85% reduction in production cycles and up to 13 parallel firings were achieved on some cycles of execution.

2.7.3 Evaluation of the Decomposition Algorithm

The decomposition algorithm is evaluated using the Manhattan Mapper expert system. Production rules of Manhattan Mapper consists of two groups, i.e., rules for man-machine interface and rules for planning. The rules for man-machine interface are allocated to the control processor, and the rules for planning (51 rules) are allocated to PEs.

The gain of parallel rule firing in this evaluation is defined as the number of production rules which can be reduced by allocating the two rules on distinct PEs. Communication costs and the effect of distributing matching processes are ignored to simplify the evaluation. Though this approximation provides a less accurate estimation of parallel firing effects, the effectiveness of the proposed decomposition algorithm can be shown based on the evaluation. More precise definition of the gain of parallel rule firing for a particular architecture, for example a distributed memory tree-structured machine, is discussed in [Ishida and Stolfo, 1984].

The evaluation is performed by gradually adding execution traces. Evaluation results on the initial partition, obtained from one execution trace, and the final partition, obtained from several additional execution traces are shown in Figure 2.11. The average number of firing rules per cycle increases by a factor of 7.57 on 32 processor systems. Though the final partition cannot be guaranteed to be the optimal partition, better partitions than the final one are difficult to obtain even by human efforts. This result shows not only that the effect of tuning is quite large, but also that the proposed incremental approach is efficient enough to develop satisfactory partitioning.

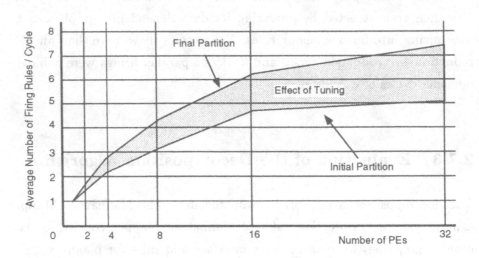

Figure 2.11: Evaluation of the Decomposition Algorithm

2.8 Related Work

Research on parallel rule matching started with the DADO project of Columbia University [Stolfo and Shaw, 1982; Stolfo, 1984], followed by the PSM project of Carnegie-Mellon University [Forgy, 1984; Gupta *et al.*, 1986]. Since the parallel rule matching does not affect the semantics of the production system programs, most of the research have focused on partitioning the RETE network and mapping the partitions to parallel processors [Miranker, 1990a; Gupta, 1987].

The parallel rule firing mechanism was first investigated by Ishida and Stolfo [1985], followed shortly by Moldovan [1986]. Unlike parallel rule matching results, results from parallel firing may become different from those of sequential firing. To cope with this problem, a data dependency approach was invented to guarantee the serializability of rule firings [Ishida and Stolfo, 1985]. Later, Schmolze [1989] and Ishida [1990] independently improved this framework, and proposed to combine compile-time and run-time analyses as described in this chapter.

Parallel rule firing has been investigated by several different universities and research institutes. Kuo and Moldovan [1992] are pursuing a problem called the *convergence problem*. Their motivation is to ensure the correctness of the parallel rule firing. Another approach is called *task-level parallelism*, which focuses on high-level decomposition of production systems [Harvey *et al*, 1991]. Implementations of parallel rule firing on hardware systems include PARULEL [Stolfo *et al.*, 1991], CREL [Kuo *et al.*, 1991] and RUBIC [Kuo and Moldovan, 1991].

A survey on parallel production systems can be found in [Kuo and Moldovan, 1992]. Recent papers on parallel rule firing are featured in the *Special Issue on the Parallel Execution of Rule Systems, Journal of Parallel and Distributed Computing* [JPDC, 1991].

2.9   Summary

A parallel firing model has been defined for production systems, and practical implementation methods have been proposed for the model: the interference analysis, the selection algorithm, the decomposition algorithm and the parallel programming environment.

The evaluation results on several production system applications show that the degree of concurrency can be increased by a factor of 2 to 9 by introducing parallel rule firing. Since the reported speed-up from parallel rule matching is 8.25 times [Gupta *et al.*, 1986], the parallel rule firing technique introduces another valuable source of parallelism to production systems.

However, a difficulty with performance evaluation is that the true number cannot be obtained from a surface analysis of existing production systems. As shown in Section 2.7, even if existing production systems do not seem to contain much parallelism, it is quite possible to optimize the systems for parallel rule firing. Careful analysis of the original problem is necessary to determine the real effectiveness of parallel rule firing.

Chapter 3

Distributed Production Systems

3.1 Introduction

This chapter first extends *parallel production systems*, in which global control exists, to *distributed production systems* with distributed control. Distributed production systems achieve speedup over synchronous parallel production systems by eliminating synchronization bottlenecks [Schmolze and Goel, 1990]. To perform domain problem solving by multiple production system agents, however, each agent needs *organizational knowledge*, which represents the necessary interactions among production system agents. This chapter formalizes the organizational knowledge as a collection of *agent-agent relationships*, which represent how agents' local decisions affect other agents' decisions.

The motive for parallel and distributed production systems is to speedup production systems several times over, and not necessarily to make them more adaptive or reactive, e.g., to follow changing environmental demands or resource constraints. Thus, the current parallel and distributed production system techniques are not yet fully adequate for *real-time expert systems* [Laffey *et al.*, 1988].

Organization self-design is thus introduced into distributed production systems to adapt itself to environmental demands: Problem solving requests issued from the environment arrive at the organization continuously, and at variable rates; to respond, the organization must adapt to chang-

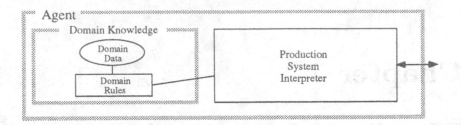

Figure 3.1: Production System Agent

ing conditions and supply meaningful results within specified time limits. However, the kind of organizational knowledge required for reorganization has not been thoroughly investigated in prior research. To achieve adaptive reorganization, this chapter extends the organizational knowledge to include *agent-organization relationships*, which represent how agents' local decisions affect the behavior of the entire organization. New reorganization primitives, *composition and decomposition of agents*, are then introduced. Composition and decomposition can occur concurrently in different parts of the organization. Through repeated application of these primitives, organization self-design can be locally and asynchronously performed.

Both composition and decomposition dynamically change the size of the agent population, the resources allocated to each agent, and the distribution of problem solving and organizational knowledge in the organization. In general, decomposition increases the overall level of resources used, while composition decreases resource use. Various simulation results show the effectiveness of the proposed organization-self design approach for building adaptive real-time systems with production system architectures.

3.2   Distributed Production Systems

3.2.1   Agent Architecture

A *production system agent* is a production system capable of interacting with other agents. As illustrated in Figure 3.1, the production system agent consists of a *production system interpreter* and *domain knowledge*, in which the PM represents *domain rules* and the WM represents *domain data*. On

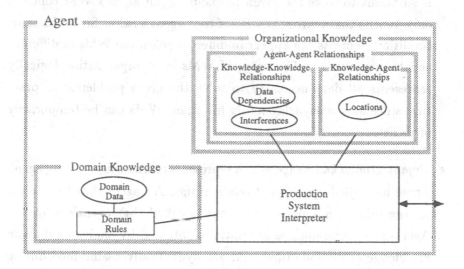

Figure 3.2: Distributed Production System Agent

the other hand, Figure 3.2 represents the architecture of *distributed production system agents*, each of which contains a *part* of the domain knowledge. Such agents must communicate with other agents for data transfer and for synchronization. Thus, each agent requires *organizational knowledge*, which represents *agent-agent relationships*. A distributed production system agent comprises the following three components:

- A *production system interpreter*, which continuously repeats the production cycle. In a parallel production system, multiple rules are simultaneously fired but globally synchronized at the Select phase as described in Chapter 2. In a collection of distributed production system agents, on the other hand, rules are asynchronously fired by distributed agents. Since no global control exists, interference among the rules is prevented by local synchronization between individual agents.

- *Domain knowledge* contained in the PM, which represents *domain rules*, and WM, which represents *domain data*. To simplify the following discussion, it is assumed that no overlap exists between PMs in different agents, and that the union of all PMs in the organization

is sufficient to solve the given problem. Each agent's WM contains
only WMEs that match the LHS of that agent's rules. Since the same
condition elements can appear in different rules, the WMs in different
agents may overlap. The union of WMs in an organization logically
represents all data necessary to solve the given problem. In practice, since agents asynchronously fire rules, WMs can be temporarily
inconsistent.

- *Organizational knowledge*, which represents relationships among agents.
  These are called *agent-agent relationships*. An agent that has such a
  relationship with a particular agent is called that agent's *neighbor*.
  Agent-agent relationships are initially obtained by analyzing domain
  knowledge at compile time, and are dynamically maintained during
  the process of organization self-design. Since agents asynchronously
  perform reorganization, organizational knowledge can be temporarily
  inconsistent across agents.

As described in Chapter 2, interference exists among rule instantiations
when the result of parallel execution of the rules is different from the results of sequential executions applied in any order. Interference must be
avoided by synchronization. Various methods for detecting interference are
discussed in Chapter 2. For distributed production systems, only compile-
time analysis is utilized because run-time analysis requires global synchronization, which is too expensive in distributed situations. In compile-time
analysis, interference can be identified when multiple rules destroy other
rules' preconditions in a cyclic fashion. (See the all-rule conditions in Chapter 2.) If the interfering two rules are distributed to different agents, the
agents have to locally synchronize to prevent the rules from being fired in
parallel and thus maintain consistency.

3.2.2   Organizational Knowledge

Agent-agent relationships can be seen as the aggregation of two more primitive types of relationships: *knowledge-knowledge relationships*, which represent interactions within domain knowledge, and *knowledge-agent relation-*

ships, which represent how domain knowledge is distributed among agents. Knowledge-knowledge relationships consist of *data dependencies* and *interferences* among domain rules as follows:

- *Data dependencies:*
  Each agent knows which domain rules in the organization have data dependency relationships with its own rules. We say that `ruleA` *depends on* `ruleB` if `ruleA` refers to a working memory node that is changed by `ruleB`. This is described as `depends(ruleA, ruleB)`. The data dependency knowledge of `agentP` is represented as:

$DEPENDENCY_{agentP}$ =
  {(ruleA, ruleB) |
    (ruleA$\in PM_{agentP}$ ∨ ruleB$\in PM_{agentP}$)
    ∧ depends(ruleA, ruleB)}

- *Interferences:*
  Each agent knows which rules in the organization may *interfere with* its own rules. The interference of `ruleA` and `ruleB` is described as `interfere(ruleA, ruleB)`. The interference knowledge of `agentP` is represented as:

$INTERFERENCE_{agentP}$ =
  {(ruleA, ruleB) |
    (ruleA$\in PM_{agentP}$ ∨ ruleB$\in PM_{agentP}$)
    ∧ interfere(ruleA, ruleB)}

Though an individual agent's execution cycle is sequential, potential interference among its own rules is analyzed for potential future distribution of those rules.

On the other hand, knowledge-agent relationships are represented by the *locations* of domain rules:

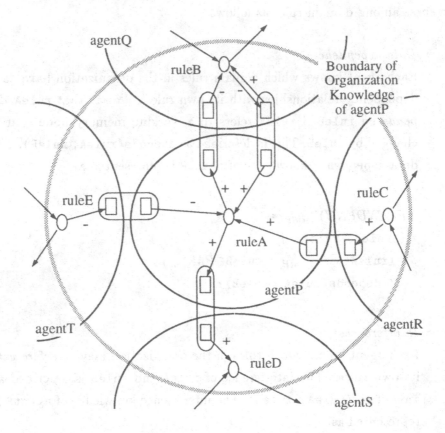

indicates the same working memory node
duplicatively stored in different agents.

$DEPENDENCY_{agentP}$ = {(ruleA, ruleB) (ruleB, ruleA)
                              (ruleA, ruleC) (ruleD, ruleA)
                              (ruleA, ruleE)}
$INTERFERENCE_{agentP}$ ={(ruleA, ruleB)}
$LOCATION_{agentP}$ = {(ruleA, agentP) (ruleB, agentQ)
                          (ruleC, agentR) (ruleD, agentS)
                          (ruleE, agentT)}

Figure 3.3: Organizational Knowledge

- *Locations:*

  Each agent, say `agentP`, knows the location of rules, say `ruleA`, appearing in its own data dependency and interference knowledge. The appearance of `ruleA` in the data dependency and interference knowledge of `agentP` is described as `appears(ruleA, agentP)`. The location knowledge of `agentP` is represented as:

  $LOCATION_{agentP}$ =
  `{(ruleA, agentQ) |`
      `appears(ruleA, agentP)` $\land$ `ruleA`$\in PM_{agentQ}$`}`

Figure 3.3 illustrates the organizational knowledge of `agentP`. Large solid circles indicate the boundaries of individual agents. Long, narrow ovals that connect agents indicate interaction paths among agents; the two rectangles within each oval indicate the WMEs communicated between agents via that interaction path, and duplicated in both agents. '+' and '-' indicate data dependencies as described in Chapter 2.

In the example in Figure 3.3, since `ruleA` and `ruleB` interfere with each other, `agentP` has to synchronize with `agentQ` when executing `ruleA`. Also, `ruleA`'s WM modification has to be transferred to `agentS`. Four agents, `agentQ`, `agentR`, `agentS`, and `agentT`, are called *neighbors* of `agentP` because they have agent-agent relationships with `agentP`. From this definition, as illustrated in Figure 3.3, `agentP`'s organizational knowledge refers only to its *neighbors*.

3.3 Distributed Firing Protocol

A *production cycle of distributed production system agents* is defined by extending the conventional *Match-Select-Act cycle* to accommodate inter-agent data transfers and synchronization. Inter-agent inconsistency caused by distribution is handled locally by using temporary synchronization via *rule deactivation*. Note that preservation of message ordering is assumed. The *distributed firing protocol* is represented in Figure 3.4.

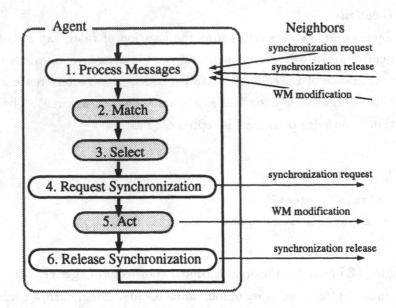

Figure 3.4: Distributed Firing Protocol

1. *Process messages:*
   When receiving a *synchronization request message* (e.g., `deactivate` `(ruleA)`) from some agent, return an *acknowledgment message* and deactivate the corresponding rule (`ruleA`) until receiving a *synchronization release message* (`activate (ruleA)`) from the same agent. When receiving a *WM modification message*, update the local WM to reflect the change made in another agent's WM.

2. *Match:*
   For each rule, determine whether the LHS matches the current WM.

3. *Select:*
   Choose one instantiation of a rule (e.g., `ruleB`) that is not deactivated.

4. *Request synchronization:*
   Using interference knowledge, send synchronization request messages (`deactivate (ruleB)`) to the agents requiring synchronization. Await acknowledgment from all synchronized agents. After complete acknowledgment, handle all WM modification messages that have ar-

rived during synchronization. If the selected instantiation is thereby canceled, send synchronization release messages and restart the production cycle.

5. *Act:*

    Fire the selected rule instantiation (`ruleB`). Using the data dependency knowledge of `agentP`, inform dependent agents with WM modification messages.

6. *Release synchronization:*

    Send synchronization release messages (`activate (ruleB)`) to all synchronized agents.

To avoid deadlock, interfering rule pairs are prioritized at compile time. This idea is borrowed from [Schmolze and Goel, 1990]. Let `ruleA` and `ruleB` interfere with each other, and let `ruleB` be given a higher priority. Then, `ruleB` can be fired without synchronization as long as it is not deactivated. However, when firing `ruleA`, `ruleB` has to be deactivated through synchronization. This approach can avoid interference through one-directional synchronization, and thus can reduce half of the synchronization overhead. Deadlock may still occur when agents are prioritized in a cyclic fashion, i.e., `ruleA` requires `ruleB` be deactivated, `ruleB` requires `ruleC` be deactivated, and `ruleC` requires `ruleA` be deactivated. However, since interference is analyzed at compile time, rules are easily prioritized such that loops are not created. Thus, this approach can avoid deadlocks among distributed production system agents.

3.4 Organization Self-Design

3.4.1 Agent Architecture

Since no single organization is appropriate in all situations, organization self-design allows an organization of problem solvers to adapt itself to dynamically changing situations. For example, suppose there are three agents in an organization, each of which fires one production rule for solving each problem request; the three agents work in a pipelined fashion (because their rules

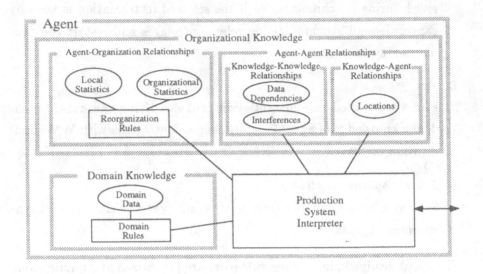

Figure 3.5: Organizational Distributed Production System Agent

are sequentially dependent), and the communication delay among agents is equal to one production cycle. Thus, the total throughput cycle time for satisfying a single request is 5. In this case, however, a single agent organization would perform better: it would incur no communication overhead, and would take only 3 cycles for satisfying a single request. On the other hand, if there were ten problem solving requests, the response time of the last request would be 14 cycles in the three-agent organization, while it would be 30 in the single agent case.

Figure 3.5 represents the architecture for *organizational distributed production system agents*. Organization self-design is performed in the following way: Upon initiation, only one agent containing all domain and organizational knowledge, exists in the organization. It is assumed that organizational knowledge for the initial agent is prepared by analyzing its domain knowledge before execution. Problem solving requests continuously arrive at the agent; older pending requests are processed with higher priority.

For appropriate reorganization, the organizational knowledge described in Section 3.2.2 is extended to include *agent-organization relationships*,

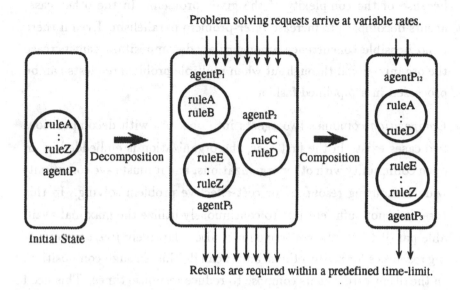

Figure 3.6: Organization Self-Design

which represent how agents' local decisions affect organizational behavior or, in other words, how well the organization is meeting its response goals. Agent-organization relationships consist of *local statistics*, *organizational statistics* and *reorganization rules*, which will be explained in Section 3.4.3.

Since the reorganization rules are also production rules, organization self-design and domain problem solving are arbitrarily interleaved. In the actual implementation, however, it is assumed that higher priority is given to the reorganization rules during the Select phase of the production cycle.

3.4.2 Reorganization Primitives

Figure 3.6 describes the process of organization self-design. For effective reorganization, agents should invoke the following reorganization primitives appropriate for each situation.

- *Decomposition* divides one agent into two. There are two cases. In the first case, agents decompose to increase *intra*-problem parallelism. This happens when the structure of the problem solving rules being applied contains concurrency, and agents cannot meet deadlines because of the complexity of the given problem. In the other case, agents decompose to increase *inter*-problem parallelism. Even if there is no possible concurrency among rules, decomposition can increase the organizational throughput when multiple problem requests can be processed in a pipelined fashion.

- *Composition* combines two agents into one. As with decomposition, two cases exist. In the first case, the organization is embedded in an open community with other organizations, and it must save community-wide computing resources for cost-effective problem solving. In this case, it is not sufficient just to continuously utilize the maximal available parallelism—the collective must also adaptively *free up* computing resources for use by others, and it can do this through composition. In the other case, agents compose to reduce response times. This need arises when communication overhead cannot be ignored. Because of the inter-agent communication overhead, maximal decomposition does not necessarily yield either minimal response time or maximal organizational throughput. Composition may actually reduce response time, even though parallelism decreases, where coordination overhead (i.e., communication and synchronization) is high.

These primitives are performed as follows:

- *Decomposition* is triggered when the environmental conditions (problem solving demand on the organization and required response-time) exceed the organization's ability to respond, given its current form and resource level. Excessive demand at the organization level is translated into excessive local demand in particular regions of the organization, measured using the local organizational statistics. At this point, particular agents with excessive local demand are divided into multiple agents, and additional computational resources are assigned to them. Decomposition continues until parallelism increases and response improves.

- *Composition* is performed when under-utilized resources can be released for use by other organizations, or to improve local performance by reducing coordination overhead. When two agents, taken together, contain an oversupply of resources, they are combined into one agent via composition. Composition repeats until no more composition is possible under the conditions of meeting deadlines.

Since the aims of composition and decomposition are independent, both kinds of reorganization can be performed simultaneously in different parts of the organization. In this way, both problem solving and organization self-design are treated as decentralized processes.

3.4.3 Organizational Knowledge

Since multiple agents asynchronously fire rules and perform reorganization, knowing the exact status of the entire organization is difficult. Under the policy of obtaining better decisions with maximal locality, *local* and *organizational statistics*, which can be easily obtained, are first introduced, and then *reorganization rules* using those statistics are provided to select an appropriate reorganization primitive when necessary.

- *Local statistics:*

  Firing ratio is introduced to represent the level of activity of each agent. Let P be a predefined period (normalized by production cycles) for measuring statistics, and F be the number of rule firings during P. Then the firing ratio, R, can be represented by F/P. When $R = 1.0$ (i.e., there are no idle production cycles over the measurement interval P), agents are called *busy*, while when $R < 1.0$, agents can be assigned additional tasks. To avoid the need for frequent communication among agents, however, it is not assumed that agents need to know other agents' local statistics.

- *Organizational statistics:*

  It is assumed that each agent can know by periodically-broadcast messages whether the organization is currently meeting deadlines. Let $T_{response}$ be the most recently observed response time (that is, time taken to complete the most recent task), and $T_{deadline}$ be the predefined time limit of the task. When $T_{response} > T_{deadline}$, the performance of the organization should be improved, while when $T_{response} < T_{deadline}$, the organization can release resources.

- *Reorganization rules:*

By using local and organizational statistics, the following rules are provided for each agent to initiate reorganization. These rules are tested during the production cycle.

R1) Perform decomposition if

$$T_{deadline} < T_{response} \text{ and}$$
$$R = 1.0$$

R2) Perform composition if

$$T_{deadline} > T_{response} \text{ and}$$
$$2R < T_{deadline}/T_{response}$$

R3) Perform composition if

$$R < 0.5$$

R1 initiates busy agents to perform decomposition, when the organization cannot meet its deadline. *R2* initiates agents to perform composition, when the organization can keep its deadline. Composition is performed even if agents are fully busy, when $T_{response}$ is enough lower than $T_{deadline}$.

R3 is introduced to take account of communication overhead. Suppose problem solving requests initially arrive frequently, and subsequently decrease. Initially, *R1* is repeatedly applied, maximizing the parallelism to increase organizational throughput. Later, even though the frequency of requests decreases, *R2* may not be valid because the communication overhead may not allow agents to meet deadlines. Thus, *R3* is necessary to merge lightly loaded agents even when $T_{response}$ exceeds $T_{deadline}$. This merging lowers coordination cost in the overall problem pipeline, and so improves performance.

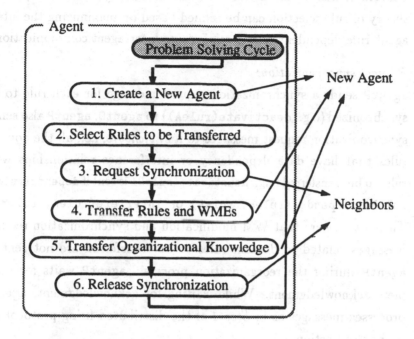

Figure 3.7: Reorganization Protocol (Decomposition)

3.5 Reorganization Protocol

Reorganization is triggered by the firing of a reorganization rule during the normal production cycle. The protocol in Figure 3.7 describes how one agent (e.g., `agentP`) decomposes itself into two agents (e.g., `agentP` and `agentQ`). Figure 3.7 also represents the decomposition protocol. During reorganization, domain rules, WMEs, dependency and interference knowledge are transferred from `agentP` to `agentQ` without any modification. However, location knowledge is modified due to the relocation of domain rules and changes are propagated to neighboring agents.

1. *Create a new agent:*

   `agentP` creates a new agent, `agentQ`, which immediately starts production cycles.

2. *Select domain rules to be transferred:*

   `agentP` selects domain rules to be transferred (e.g., `ruleA`) to `agentQ`.

Currently, half of the active rules are arbitrarily transferred, but a theory of rule selection can be refined based on maximizing the intra-agent rule dependencies and minimizing inter-agent communication.

3. *Request synchronization:*
   `agentP` sends a synchronization request message for each rule to be synchronized (e.g., `deactivate(ruleA)`) to `agentQ`. `agentP` also sends synchronization request messages to its *neighbors,* i.e., all the domain rules that have data dependency or interference relationships with rules to be transferred (e.g., `deactivate(ruleB)` is sent if `depends(ruleA, ruleB)`, `depends (ruleB, ruleA)` or `interfere(ruleA, ruleB)`.) This is to assure that WM modification and synchronization request messages related to domain rules to be transferred are not sent to `agentP` during the reorganization process. `agentP` waits for complete acknowledgment. While waiting for acknowledgment, `agentP` processes messages as in *Step 1* of the distributed firing protocol described in Section 3.3.

4. *Transfer rules:*
   `agentP` transfers rules (`ruleA`) to `agentQ`, updates its own location knowledge, and propagates the change to its *neighbors.*

5. *Transfer WMEs:*
   `agentP` copies WMEs that match the LHS of the transferred rules (`ruleA`) to `agentQ`. More precisely, to avoid reproducing once-fired instantiations, not only WMEs but also conflict sets are transferred to `agentQ`. Before transferring the conflict sets, however, `agentP` has to maintain its WM by handling the WM modification messages that have arrived before the synchronization is completed. A bookkeeping process follows in both agents to eliminate duplicated or unneeded WMEs.

6. *Transfer dependency and interference knowledge:*
   `agentP` copies its dependency and interference knowledge to `agentQ`. Both agents do bookkeeping to eliminate duplicated or unneeded organizational knowledge. Unneeded data dependency and interference

knowledge are tuples that include none of the agents' rules. Unneeded location knowledge consists of tuples that include none of the rules that appear in the agents' data dependency and interference knowledge.

7. *Release synchronization:*

agentP sends synchronization release messages (activate(ruleA) to agentQ, and activate (ruleB) to all synchronized *neighbors*). This ends reorganization.

An agent (e.g., agentP) can compose with another agent by a similar process. First, agentP sends *composition request messages* to its *neighbors*. If some agent, say agentQ, acknowledges, agentP transfers all domain and organizational knowledge to agentQ and destroys itself. The transfer method is the same as that for decomposition.

During the reorganization process, deadlock never occurs, because reorganization does not block other agents' domain problem solving and reorganization. Furthermore, though neighboring agents are required to deactivate domain rules that depend on or interfere with the transferred rules, they can concurrently perform other activities including firing and transferring rules that are not deactivated. This localization helps agents to modify the organization incrementally.

3.6 Evaluation Results

To evaluate the effectiveness of the proposed approach, a simulation environment is implemented and the *Waltz labeling program* is executed: 36 rules solve the problem that appears in Figure 3-17 in [Winston, 1977] with 80 rule firings. The complete program is included in Appendix B. Experiments begin with one agent that contains all problem solving knowledge. Its organizational knowledge is trivial in that references are to itself, since it has no neighboring agents.

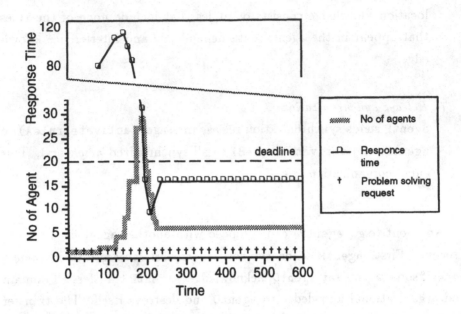

Figure 3.8: Evaluation with Constant Intervals

3.6.1   Simulation Excluding Overheads

Figures 3.8 and 3.9 show the simulation results. In these figures, communication and reorganization overheads are ignored. The line chart indicates
response times normalized by production cycles. The step chart represents
the number of agents in the organization. The time limit ($T_{deadline}$) is set
at 20 production cycles, while the statistics measuring period (P) is set
at 10 production cycles. In Figure 3.8, problem solving requests arrive at
constant intervals, while in Figure 3.9, the frequency of requests is changed
periodically. From these figures, the followings are concluded:

- *Adaptiveness of the organization:*
  In Figure 3.8, around time 100, the response time far exceeds the time
  limit. Thus the organization starts decomposition. Around time 200,
  the number of agents has increased to 26, the response time drops
  below the time limit, and the organization starts composition. After
  fluctuating slightly, the organization finally reaches a stable state with

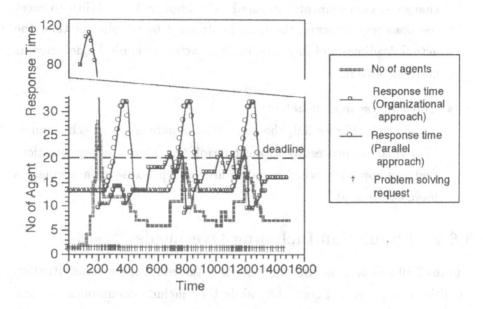

Figure 3.9: Evaluation with Changed Intervals

the number of agents settling at 6. Since composition and decomposition have been repeatedly performed, the firing ratios of the resulting agents are almost equal. In Figure 3.9, the number of agents at the busiest peak decreases over time. Both charts show that the society of agents has gradually adapted to the situation through repeated reorganization.

- *Real-time problem solving:*
The average number of agents in Figure 3.9 is approximately 9. Response times of the organizational approach, which flexibly selects the number of agents, are compared to those of the *conventional parallel approach* using 9 permanent agents. Differences in results from these two approaches demonstrate that while the conventional approach uses the same average number of agents, it cannot respond to meet deadlines when problem demand increases. Thus, the organizational approach is more effective for adaptive real-time problem

solving. However, the effect of reorganization does lag behind the change in environmental demand. For improved capability to meet response requirements, the time limits must be set shorter than the actual deadlines and increases in agent activity should be detected as early as possible.

- *Efficient resource utilization:*
  As shown in Figure 3.9, the conventional parallel approach requires 17 permanent processors to meet deadlines. Thus, the organization-centered approach, which requires around 9 processors on average, is more economical.

3.6.2   Simulation Including Overheads

Figures 3.10 and 3.11 describe the results obtained from the same situation conditions as given in Figure 3.8, while they include communication and reorganization overheads. What are reasonable assumptions for communication and computation speeds? In the iPSC/2, a typical message-passing multicomputer, the communication overhead of sending a $2Kbyte$ message across the diameter of a 128 node machine is $840\mu sec$, and the transfer rate for a $64Kbyte$ message is $2.6Mbyte/sec$ [Bomans and Roose, 1989]. For computation, a state-of-the-art production system takes from several to several tens of $msec$ for one production cycle on an HP9000/370, a Motorola 68030 based workstation [Miranker *et al.*, 1990]. Since one production cycle creates several messages, each of which contains a few WMEs, the communication overhead in a good message-passing machine can be estimated as at most one production cycle. However, communication overhead have been taken into account to cover cases in which wider-area and somewhat slower networks such as *Ethernet* or *public telecommunication networks* are used for distributed problem solving. Let O_c be the average network latency represented in terms of production cycles. Then, agents can utilize other agents' results no sooner than O_c cycles later. Simulations have been performed on situations in which O_c was equivalent to 1, 3 or 5 production cycles to assess the effect of communication overheads.

Reorganization overheads cannot be ignored even in message passing

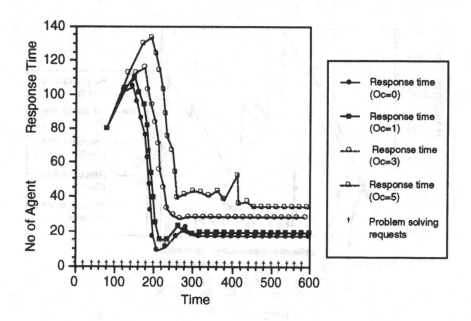

Figure 3.10: Evaluation with Communication Overheads

machines, depending on how many rules, WMEs, and conflict sets are to be transferred. When the RETE match algorithm [Forgy, 1982] is employed, building the RETE networks in newly generated agents requires additional costs. However, this can be ignored when assuming the TREAT match algorithm [Miranker, 1987], in which networks are built dynamically in each production cycle. Let O_r be the reorganization overhead in terms of production cycles. O_r of the example program costs at most 10 production cycles, during which all rules of the Waltz labeling program and WMEs for 10 pending problem solving requests can be transferred. However, simulations have been done on cases where O_r is equivalent to 10, 30 or 50 production cycles to observe the general influence of reorganization overheads on organization self-design for distributed production systems. The major results obtained from these simulations are as follows:

- *Influences of communication overhead:*
  Figure 3.10 considers communication overhead but does not include reorganization overhead. When $O_c = 1$, the organization can meet

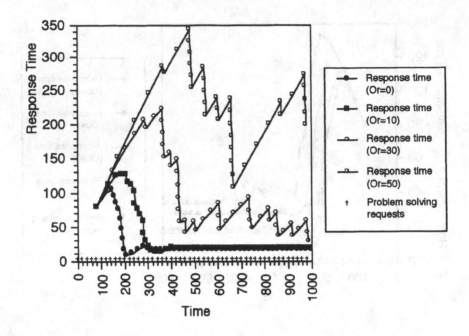

Figure 3.11: Evaluation with Reorganization Overheads

its deadline, but when $O_c = 3$ or more, the organization fails to satisfy the real-time constraint. This is because communication overhead delays problem solving, and this destablizes the organization. The organization fluctuates in two cases. Agents may decompose themselves rapidly so that $T_{response}$ becomes much less than $T_{deadline}$. This triggers $R2$, and causes agents to start composition. The other case occurs even when $T_{response}$ exceeds $T_{deadline}$, i.e., the firing ratio of agents significantly decreases because of the communication overheads. In this case, $R3$ is satisfied. The chances of the latter case increase with the communication overhead.

- *Influences of reorganization overhead:*
  Figure 3.11 considers reorganization overhead but does not include communication overhead. Unlike the communication overhead, reorganization overhead is temporary and thus should not affect the stability of the organization. When $O_r = 10$, the organization soon

reaches a stable state. However, when reorganization overhead becomes larger, such as $O_r = 30$ or more, the organization oscillates and never seems to become stable. The reason is as follows. Since reorganizing agents cannot fire rules during the decomposition process, their firing ratios temporarily decrease. Firing ratios of neighboring agents also decrease because no new WME is transferred from the reorganizing agents. As a result, $R3$ is fired in the neighboring agents to start composition, and thus the organization oscillates.

Communication overhead is not a problem in current message passing machines. Furthermore, ongoing research on message passing machines has been reduced the communication overhead by an order of magnitude [Dally, 1986]. However, in the future, communication overhead can be a problem when using wider networks to perform distributed problem solving. Reorganization overhead is also not a problem in this example, but it too might cause oscillation if it is too large. Further research is required, but one way to avoid oscillation due to reorganization would be to decrease the sensitivity of reorganization by enlarging the period of measuring statistics (P).

3.7 Related Work

The research on distributed production systems can be classified into two categories: *distributed rule matching*, where the RETE network is partitioned and distributed into multiple processes, and *distributed rule firing*, where rules are distributed as described in this chapter. Hsu *et al.* [1987] investigated distributed rule matching. Since the motivation of distributed rule matching is somewhat unclear, this technique has not been widely accepted. On the other hand, Schmolze and Ishida independently studied distributed rule firing to eliminate synchronization overheads in parallel rule firing [Schmolze and Goel, 1990; Ishida *et al.*, 1990].

The organization self-design approach proposed in this chapter can complement other approaches currently being developed for real-time expert systems, such as *approximate processing techniques* [Lesser *et al.*, 1988] and *adaptive intelligent systems* [Hayes-Roth *et al.*, 1989]. These approaches attempt to meet deadlines by improving the decision-making of individual agents. On the other hand, the organization self-design approach, where problems are solved by a society of distributed problem solving agents, aims to achieve adaptive real-time performance through the reorganization of the society.

The performance of *rigid* organization structures was studied by Malone [1987]. Various problem solving organizations were investigated by Corkill [1982]: the load balancing of spatially distributed agents, for example, was studied to efficiently track vehicles crossing over areas with multiple agents. However, no mechanisms for organization self-design have yet been implemented, and therefore, the dynamic behavior of problem solving organizations has not actually been observed. Furthermore, previous research on reorganization typically aimed at changing agent roles or inter-agent task ordering [Corkill, 1982; Davis *et al.*, 1983; Durfee and Lesser, 1987]. On the other hand, this chapter introduced new reorganization primitives, composition and decomposition of agents, and reported an actual performance of reorganization in real-time.

Hogg and Huberman [1990] have studied oscillation problems similar to those observed in Figure 3.11. They verified the possibility of chaotic behavior in systems with long communication delays, and suggested an approach to controlling chaos based on rewarding agents with good decision-making performance. However, their scheme fixes both the boundary and the decision capability of an agent, whereas in the proposed formulation, an agent is a flexible entity, and it is less clear where to assign credit or blame for poor performance over the longer term.

All organization self-design studies until now have assumed inherently decomposable systems, such as blackboard or production systems consisting of independent knowledge components. Further research is required to apply organization self-design to more general cases of problem solving by multiple agents, such as multiple robot systems, which include actions, plans and conflicts among agents. The initial report along this direction can be found in [Ishida, 1993].

3.8 Summary

Techniques for building problem solving systems capable of adapting to changing environmental conditions are of great interest. This chapter has presented an approach that relies on reorganization of a collection of problem-solvers to track changes in deadlines and problem solving requests. The approach exploits an adaptive trade-off of resources and organization form to satisfy time and performance constraints. Agents are created and destroyed, and domain knowledge is continually reallocated. To extend the possible architectures for organization self-design, composition and decomposition have been introduced as new reorganization primitives. Organizational knowledge has been formalized to represent interactions among agents and their organization.

With additional decision-making meta-knowledge, this approach can become a more general organization self-design technique. It also has the advantage of being grounded in a well-understood body of theory and practice: parallel production systems. In the current version, composition/decomposition decisions are made solely on the basis of firing ratios, and the choice of rules to transfer is made arbitrarily. Note, however, that allocation decisions could instead be based on the semantics of rules (i.e., distribution based on the kinds of tasks that need more resources). Moreover, partial knowledge transfer among *existing* agents can be combined with the composition and decomposition to provide a flexible and distributed task-sharing system.

Chapter 4

Multiagent Production Systems

4.1 Introduction

The previous two chapters describe two types of concurrent production systems: synchronous parallel production systems or parallel rule firing systems, where rule firings are globally synchronized in each production cycle, and asynchronous parallel production systems or distributed production systems, where the rules are fired in parallel without global synchronization. Most of the above research, however, aimed at improving the performance of a single production system program. This chapter extends the theory of parallel and distributed production systems to create a basis for *multiagent production systems*, where multiple production system programs compete or cooperate to solve a single problem or multiple problems. Shared information architectures will be focused on, where multiple agents interact through shared working memory elements. This architecture has recently been attracting attention as a way of implementing the following problem solving paradigms.

Data-oriented problem solving, such as case-based problem solving, which is becoming widely accepted as a complement to rule-oriented problem solving. A database approach to production systems has also been investigated in [Raschid *et al.*, 1988; Miranker, 1990a; Sellie and Lin, 1990], where the working memory is not merely a short-term memory, but also contains sharable persistent data to be concurrently referenced and changed by multiple problem solving systems.

Cooperative distributed problem solving, which is a paradigm for multiple autonomous agents to cooperatively solve a single problem. Maintaining consistency among multiple agents has been discussed as a major research issue. A shared information architecture, which is examined in this chapter, can be an alternative approach to maintaining consistency among multiple agents.

The research issue addressed in this chapter is how to provide a transaction model that guarantees the consistency of a shared working memory. A new transaction model will be introduced, which is well suited to multiagent production systems. For an experimental evaluation of the proposed model, the *Monkey & Bananas* program [Brownston *et al.*, 1985], where a single monkey seeks some bananas, will be extended to *Multiple Monkeys & Bananas*, where more than two monkeys compete or cooperate with each other to get bananas.

4.2   Multiagent Architecture

4.2.1   Shared or Distributed Working Memory

Figure 4.1 provides an overview of multiagent production system architectures. A *shared-memory multiagent production system* is shown in Figure 4.1(a), where multiple production system interpreters refer to their own PMs and to one *shared WM*. A *transaction manager* executes transactions concurrently issued by the multiple production system interpreters. On the other hand, Figure 4.1(b) shows a *distributed-memory multiagent production system*, where shared information is distributed among different WMs. In this case, transaction managers provide an integrated view of distributed WMs, i.e., each production system agent does not have to know where a particular WME is physically located. The consistency of distributed WMs is maintained through cooperation among multiple transaction managers.

In both architectures, transaction managers have to guarantee the *serializability* of transactions, and maintain the *consistency* of the shared WM.

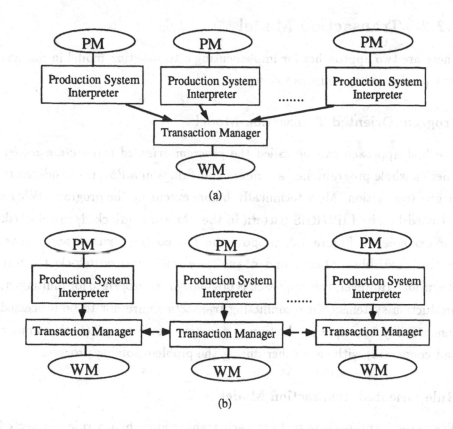

Figure 4.1: Multiagent Production System Architectures

Furthermore, as described in Chapter 1, each production system interpreter holds information about the shared WM in its discrimination network, called the RETE network. Therefore, the transaction manager is requested to report the WM changes, which are made by any production system interpreter, to all other production system interpreters to update the information held by them.

Though the multiagent production system architectures share various problems with distributed database systems, this chapter is intended to solve problems peculiar to multiagent production systems, i.e., the shared-memory architecture will be focused on, because the distributed-memory architecture can be easily obtained by applying distributed database techniques to the shared-memory architecture.

4.2.2   Transaction Model

There are two approaches for implementing a transaction model in multiagent production systems as follows.

Program-Oriented Transaction Model

The first approach can be called the *program-oriented transaction model,* where a whole program, i.e., a series of rule firings in a PM, is considered to be one transaction. More technically, before executing the program, WMEs matched by any LHS/RHS pattern in the PM are readlocked/writelocked. For example, in Figure 4.2, suppose the PM contains `ruleA` and `ruleB`. In this case, both `class1` and `class2` are read- and writelocked before execution. However, this approach is obviously inappropriate for multiagent production systems: Since conflicting transactions are not to be processed concurrently, multiple production systems cannot interact (both compete and cooperate) with each other during the problem solving process.

Rule-Oriented Transaction Model

The second approach is to form each transaction when a rule is selected for firing, that is, only WMEs matched to LHS/RHS patterns in the firing rule are readlocked/writelocked. This is called the *rule-oriented transaction model.* For example, in Figure 4.2, before executing `ruleA`, `class1` is readlocked and `class2` is writelocked. Thus, the range of locked WMEs is significantly restricted. Furthermore, since WMEs are always unlocked after rule firing compilation, this approach allows multiple production systems to interact with each other. Therefore, the rule-oriented transaction model is appropriate for multiagent production systems. However, this model forces us to face the following two problems.

Concurrency control problem:

> Since transactions are concurrently created by multiple production system interpreters, the transaction manager is required to guarantee the serializability of multiagent transactions. Furthermore, since production system interpreters contain information about the shared

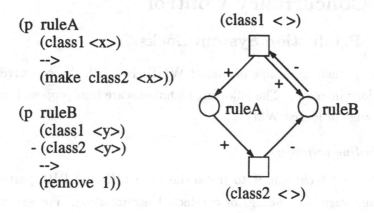

(p ruleA
  (class1 <x>)
  -->
  (make class2 <x>))

(p ruleB
  (class1 <y>)
  - (class2 <y>)
  -->
  (remove 1))

Figure 4.2: Example Data Dependency Graph

WM, the transaction manager should also guarantee the consistency of the shared information distributed among multiple production system interpreters. An efficient concurrency control protocol called the *lazy lock protocol*, will be introduced in Section 4.3 to cope with this problem.

Inter-rule consistency problem:

In the rule-oriented transaction model, each transaction is associated with one rule firing. *A transaction is a unit of consistency, that is, if executed without interference from other transactions, it transforms a database from one consistent state to another* [Bernstein and Goodman, 1981]. However, since a single rule is not always a unit of consistency, the shared WM could become inconsistent with interleaved rule firings. A *logical dependency model* of the shared WM and its *maintenance mechanisms* will be introduced in Section 4.4 to solve this problem.

4.3   Concurrency Control

4.3.1   Production System Locks

In general, when the range of locked WMEs is small, the concurrency of transactions increases. The following techniques are thus proposed to localize the range of locked WMEs.

Instantiating patterns:

The first technique is to instantiate the LHS and RHS patterns by using variable bindings of a selected instantiation. For example, in Figure 4.2, if `ruleB`'s instantiation is selected to be fired and if its pattern variables are bound as $\{< y >\} = \{10\}$, then the WMEs matched to (`class1` 10) and (`class2` 10) are readlocked instead of (`class1` <>) and (`class2` <>). Similarly, (`class1` 10) is writelocked instead of (`class1` <>).

Introducing a new lock system:

Table 4.1(a) represents the combinations of locks and their conflicting locks used in database systems. Note that locks requested from the same transaction are assumed not to conflict with each other. These locks are established in the following way. Data should be readlocked/writelocked before being referenced/changed. When data is readlocked, writelock is inhibited. When writelocked, both readlock and writelock are inhibited. However, database locks are not efficient in production systems, because the pipeline parallelism cannot be utilized. For example, suppose a rule to add WMEs is selected to be fired. Then, a set of WMEs that matches the RHS patterns of the rule is writelocked. Subsequently, this prevents other production systems from firing rules which refer to the writelocked WMEs. As a result, production systems cannot fire rules simultaneously in a pipeline fashion. Chapter 2 shows that pipeline parallelism has a considerable effect on the efficiency of parallel production systems.

Table 4.1(b) represents a new lock system efficient for production systems: readlocks are classified into '+'referlock and '-'referlock, and

Current Lock	Conflicting Locks
readlock	writelock
writelock	readlock, writelock

(a) Database Systems

Current Lock	Conflicting Locks
'+'referlock	'-'changelock
'-'referlock	'+'changelock
'+'changelock	'-'referlock, '-'changelock
'-'changelock	'+'referlock, '+'changelock

(b) Production Systems

Table 4.1: Conflicting Locks

writelocks into '+'changelock and '-'changelock. Conflicting locks
are derived from the *paired-rule conditions* in Chapter 2. Using this
classification, for example, even when '+'changelocked, '+'referlock
becomes to be permitted. Therefore, multiple production systems
can share the same set of WMEs, and fire rules simultaneously in a
pipeline fashion.

4.3.2 Lazy Lock Protocol

The most straightforward way to utilize the production system locks is to
establish locks before starting each production cycle. However, production
systems refer to a wide range of information due to its non-procedural programming style. All WMEs matched to all LHS and RHS patterns in all
rules should be locked. In practice, however, only a few working memory
elements are referenced by the instantiation selected to be fired. Therefore,
in this way, the range of the locked WMEs is far wider than the range of
WMEs actually referenced and changed by the selected instantiation.

To efficiently guarantee the serializability of multiagent transactions, the
lazy lock protocol is thus introduced, which consists of the following mes-

sages. The message order is assumed to be preserved during communication.

1. *Submitting a transaction:*

   Submit(transaction-ID, transaction)

   When the production system interpreter starts executing the RHS
   of a selected rule, a Submit message is issued from the production
   system interpreter to the transaction manager. Upon receiving the
   Submit message, the transaction manager investigates the contents
   of the specified transaction and tries to establish the necessary locks.
   The transaction manager processes Submit messages one by one (thus
   deadlock does never occur), and performs the *two phase lock protocol*
   [Bernstein and Goodman, 1981] (thus the serializability of transactions is guaranteed). If successful, the transaction manager executes
   the transaction, modifies the contents of the shared WM, then reports
   the WM changes to all related production system interpreters. However, the transaction manager may fail to secure the necessary locks,
   if they conflict with currently active locks. In this case, the transaction manager enqueues the transaction until the conflicting locks to
   be released.

2. *Reflecting WM changes:*

   Reflect(transaction-ID, WM-changes)

   A Reflect message is issued from the transaction manager to all
   related production system interpreters, when the shared WM is updated. A Reflect message triggers the production system interpreters
   to perform the Match phase, that is, the information held in the
   production system interpreters is maintained by reflecting the WM
   changes. The conflict set of the production system interpreter is also
   updated.

3. *Aborting a transaction:*

   Abort(transaction-ID)

   As a result of updating the conflict set, the currently selected instantiation might be canceled. In this case, an Abort message is issued
   from the production system interpreter to the transaction manager to

terminate the previously submitted transaction. The cancellation of the currently selected instantiation is ignored when it is caused by the instantiation itself. Note that the transaction to be aborted cannot yet have been executed, because its necessary locks would, at least, conflict with the transaction triggering the cancellation.

4. *Completing the reflection of WM changes:*
Done(transaction-ID)
After reflecting the WM changes in the information held in the production system interpreter, a Done message is issued from the production system interpreter to the transaction manager. When receiving Done messages from all related production system interpreters, the transaction manager releases the locks requested by the specified transaction.

5. *Ending a transaction:*
End(transaction-ID)
An End message is issued from the transaction manager to the production system interpreter, to finish the specified transaction. There are two cases: When the transaction is completed, the End message is issued after releasing locks. On the other hand, when the transaction is aborted, an End message is issued after removing the transaction from the waiting queue. The End message triggers the production system interpreter to resolve any conflict and to select an instantiation to be executed in the next production cycle.

Figures 4.3 and 4.4 represent an example message sequence of the *lazy lock protocol.* The major characteristics of this protocol observed from these figures are as follows:

- A transaction is created after conflict resolution, i.e., by the time the transaction is submitted, the WMEs have already been referenced by the selected instantiation. This is why we call this protocol the *lazy lock protocol.*

- In this protocol, a production cycle starts upon receipt of an End message and ends upon receipt of the next End message. Note that, as in

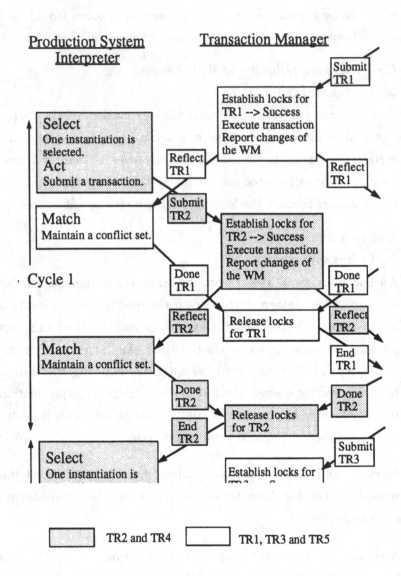

Figure 4.3: Example of Lazy Lock Protocol (1)

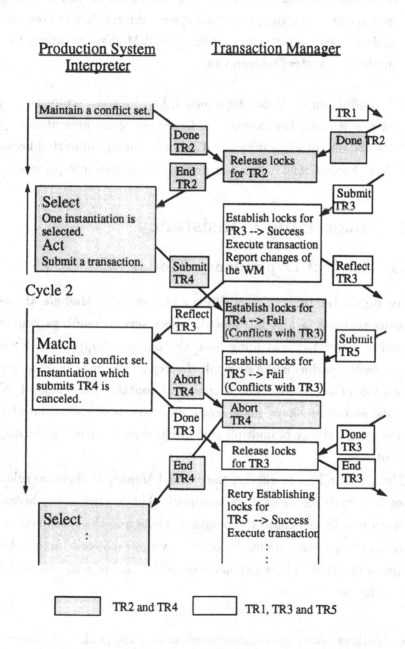

Figure 4.4: Example of Lazy Lock Protocol (2)

the first cycle presented in Figure 4.3, a production cycle can contain more than one Match phase. This is because, even in the middle of a production cycle, the production system interpreters are requested to update their conflict sets to reflect the WM changes caused by other production system interpreters.

- A production cycle can be aborted, i.e., in some production cycles, no rule is fired. For example, in the second cycle presented in Figure 4.4, an instantiation is canceled just after being submitted because of WM changes caused by other production system interpreters.

4.4   Inter-Rule Consistency

4.4.1   Logical Dependency Model

To investigate the inter-rule consistency problem, the *Multiple Monkeys & Bananas* example is first introduced, where several monkeys compete or cooperate to get bananas hung from the ceiling. Suppose each monkey independently performs the original *Monkey & Bananas* rules as given in [Brownston *et al.*, 1985]. The shared WM contains two kinds of WMEs, room status and monkeys' plans (each plan consists of collections of goals), and the serializability of multiple rule firings is guaranteed by the lazy lock protocol.

The question here is whether the original *Monkey & Bananas* rules work or not in a multiple monkey environment. The answer is *no*, because interleaved rule firings cause the monkeys' plans and the room status to be inconsistent: the plans are made obsolete by other monkeys' actions but still remain in the WM. These two kinds of WMEs are related but in different ways as follows.

- *Monkeys' goals* are created based on existing goals and current room status. These dependencies are *logical*, that is, when either the existing goals or the room status change, the created goals become obsolete and need to be retracted.

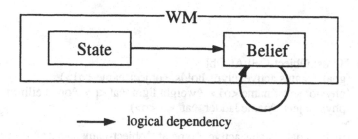

Figure 4.5: Logical Dependency Model

- *Room status* is changed based on the monkeys' goals and previous room status. These dependencies are *not logical*, that is, once the room status is changed because of a monkey's goal, it cannot be reinstated even if the monkey's goal is retracted. To reinstate the room status (e.g., to put a ladder back to its original position), another action (e.g., to move the ladder) is required.

In the following discussion, WMEs are thus classified into two types: *belief* which logically depends on other WMEs, and *state* which does not. Figure 4.5 represents the possible logical dependencies among *belief* and *state* WMEs.

4.4.2 Maintenance Mechanism

Mechanisms for maintaining logical dependencies is now introduced. Figure 4.6(a) represents some of the *Monkey & Bananas* rules. Suppose a WM class **goal** is declared to be *belief*, and **monkey** and **phys-object** to be *state*. Figure 4.6(b) illustrates the logical dependencies among WMEs, which are created by executing the rules given in Figure 4.6(a). In Figure 4.6(b), a data dependency graph is utilized to represent logical dependencies. The W-node represents a set of WMEs to be locked or unlocked. The mechanisms for maintaining logical dependencies are as follows.

Creating logical dependencies:

> Logical dependencies are created only when the newly created WMEs are classified into *belief*.

```
(p Holds::Object-Ceil:At-Obj
    (goal ^status active ^type holds ^object-name <o1>)
    (phys-object ^name <o1> ^weight light ^at <p> ^on  ceiling)
    (phys-object ^name ladder ^at <> <p>)
 -->
    (make goal ^status active ^type at ^object-name ladder ^to <p>))

(p At::Object:Holds
    (goal ^status active ^type at ^object-name <o1> ^to <p>)
    (phys-object ^name <o1> ^weight light ^at <> <p>)
    (monkey ^holds <> <o1>)
 -->
    (make  goal ^status active ^type holds ^object-name <o1>))
```

(a) Production Rules

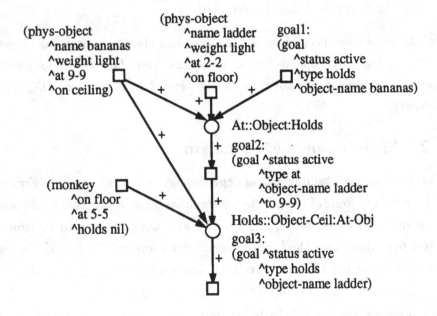

(b) Logical Dependencies to be Maintained

Figure 4.6: Maintaining Logical Dependencies

- Set a positive link (labeled '+') from a WME, which is matched by the LHS pattern of a fired rule, to the created *belief* WME. In this case, the *belief* WME is positively supported by the matched WME.

- Instantiate the negative LHS patterns of a fired rule by using the variable bindings of the instantiation. Set a negative link (labeled '-') from the instantiated pattern to the *belief* WME. In this case, the *belief* WME is negatively supported by the instantiated pattern.

Removing logical dependencies:

Logical dependencies are removed when WM changes occur. Repeat the following until no more change occurs in the WM.

- When a WME is removed, remove any WME which is positively supported by the removed WME.

- When a new WME is created, remove any WME which is negatively supported by the pattern matched to the created WME.

For example, in Figure 4.6(b), if some **phys-object** is relocated by other monkeys, **goal2** and thus **goal3** are removed. As a result, a monkey must reconstruct its plan using both **goal1** and the new room status.

Locking dependent WMEs:

Introducing logical dependencies requires us to extend the range of WMEs to be locked. When establishing locks, therefore, the following operations are to be recursively performed.

- Prevent any other production system from inappropriately changing any WMEs which support WMEs referenced by a production system. When establishing a '+'referlock to a W-node, say W_A, establish '+'referlocks to all W-nodes that positively support W_A, and establish '-'referlocks to all W-nodes that negatively support W_A.

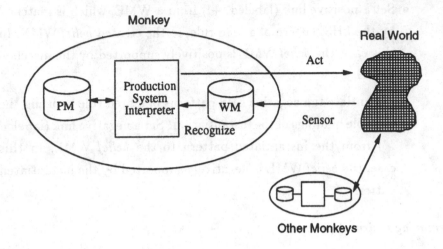

Figure 4.7: Multiple Monkeys and Bananas

- Prevent any other production system from inappropriately refer-
  encing any WMEs supported by WMEs changed by a production
  system. When establishing a '-'changelock to a W-node, say W_A,
  establish '-'changelocks to all W-nodes that are positively supported by W_A. When establishing a '+'changelock to W_A, establish '-'changelocks to all W-nodes that are negatively supported
  by W_A.

For example, in Figure 4.6(b), when applying a '+'referlock to goal3,
all other patterns in this figure are to be '+'referlocked.

4.5 Evaluation Results

Shared-memory multiagent production systems have been implemented on
the Genera multiprocess environment of Symbolics workstations and the
Multiple Monkeys & Bananas problem has been experimentally solved. The
system configuration is illustrated in Figure 4.7.

It has been observed that interference among multiple monkeys is prevented by the lazy lock protocol. In the real world, only one monkey can
grasp an object at any one time: even though the delay may be small, no

other monkey can simultaneously grasp the same object. In other words, the real world naturally serializes conflicting transactions. By way of contrast, in our experience, interference among monkeys is successfully prevented by the lazy lock protocol.

In the experiment of parallelizing Hearsay-II [Fennell and Lesser, 1981], the lock overhead was reported to be 100%. In the lazy lock protocol, since WMEs to be locked are represented by patterns, more overhead is required to examine conflicts between patterns to be locked and currently locked. Though there is room to improve the efficiency of the current implementation, a fundamental tradeoff exists between the following two approaches: executing transactions sequentially, and executing transactions concurrently while guaranteeing serializability via locks. The sequential approach might be more efficient if the rules only modify a relatively small WM. Note that when the transaction manager executes transactions sequentially, locks are no longer necessary. However, messages introduced in Section 4.3 are still effective. This is because, regardless of how the transaction manager works, multiple production systems independently perform their rules, and thus submit transactions simultaneously. Furthermore, if the WM is stored in databases or each rule firing takes considerable time, such as when controlling a robot in the physical world, the merit of concurrent execution may exceed the lock overhead.

4.6 Related Work

To maintain consistency among multiple agents, various approaches such as the *Distributed Truth Maintenance System (DTMS)* [Bridgeland and Huhns, 1990], *Distributed Assumption-based Truth Maintenance System (DATMS)* [Mason and Johnson, 1989] and *Distributed Constraint Satisfaction Problems (DCSP)* [Yokoo et al., 1992] have been investigated. A shared information architecture is an alternative approach to maintaining consistency among multiple agents.

Though this chapter only discusses the serializability of multiagent rule firings, the semantics of multiagent plan interactions could be investigated more precisely. For example, multiagent plan synchronization for deadlock avoidance has appeared in [Georgeff, 1983; Stuart, 1985], and various interactions between multiagent plans have been classified in [von Martial, 1989].

For concurrency control, related work has been done in the area of database systems [Bernstein and Goodman, 1981] and recently applied to blackboard systems [Ensor and Gabbe, 1985].

4.7 Summary

This chapter has proposed a transaction model that enables multiagent production systems to interact with each other by referring to or changing shared WM information. A database transaction model cannot be applied for this purpose, because it basically aims to enable multiple processes to be executed independently rather than interactively. In the proposed transaction model, a transaction is formed each time a rule is selected for firing. An efficient concurrency control protocol called the *lazy lock protocol* is proposed that guarantees inter-agent consistency.

However, as a result of allowing interleaving of rule firings among multiagent production systems, the inter-transaction consistency becomes a new problem. One solution is presented to overcome this problem: to formalize a logical dependency model for the shared WM, and maintain those dependencies each time the WM is updated. Further study is required to guarantee inter-transaction consistency and so extend the applicability of multiagent production systems.

Chapter 5

Meta-Level Control of Multiagent Systems

5.1 Introduction

This chapter explores multiagent production systems in more open and dynamic environments. The challenge is to analyze the complex system operation tasks performed by human operators and duplicate them on multiagent production systems. Typical complex systems are communication networks, which are controlled by cooperative multiagent operation plans usually effected by several human operators. For example, fault repairing processes are performed by groups of professionals placed in neighboring *network operation centers* . Telecommunication agencies have developed a large set of such cooperative plans, which have the following characteristics:

- Because of the variety of network states, operation plans cannot be fully specified in advance. Thus, the plans must be adapted to the dynamically changing real world at execution time.

- Various events occur asynchronously in the network; therefore, agents are required to concurrently execute multiple operation plans, and to cooperatively control the execution of their plans.

In order to control multiagent rule firings, this chapter first introduces meta-level control of production systems as follows.

- Production systems are be viewed as a collection of concurrent rule processes: each process continuously monitors the global database, and always tries to execute actions when its conditions match database entries. The *Procedural Control Macros (PCMs)*, which are based on Hoare's *CSP (Communicating Sequential Processes)* [Hoare, 1987], are thus introduced to enable control plans to adaptively communicate with production rules. Since PCMs include nondeterministic properties, the execution order of rules is not necessarily determined in advance, but can be guided by the PCMs at run-time.

- Rather than creating a new control plan description language, the approach is taken to allow existing procedural languages to control production systems. The PCMs are thus embedded into procedural programming languages. Though the PCMs are functionally simple and easy to implement, they can effectively control production systems when combined with the original control facilities of the procedural programming languages.

Based on the above meta-level control facilities, a multiagent system called *CoCo* has been implemented, which aims to describe and concurrently execute cooperative operation plans. The CoCo agent consists of a *database*, an *action interpreter*, and a *plan interpreter*. The database represents the real world. Actions are performed by the action interpreter when their preconditions are satisfied by the current database contents and desired by *operation plans*. A *meta-plan* is performed by the plan interpreter to cooperatively focus multiple agents' attention on a particular operation plan. While CoCo is suitable for several areas, this chapter concentrates on *PSTN (public switched telephone network) operations*. In the following sections, a model of action selection, a plan description language, its execution mechanism, and a method of monitoring multiagent interactions are to be described.

5.2 Conventional Control Techniques

5.2.1 Explicit and Implicit Control

In production systems, multiple rules communicate with each other through the WM. While this model is well suited to represent independent chunks of declarative knowledge, various coding tricks are required when implementing procedural control transfers from rules to rules.

The execution control of production systems can be classified into two categories: *conflict resolution*, which determines the search strategies but does not affect the solution set, and *procedural control*, which constrains production invocation and thereby modifies the solution set [Georgeff, 1982]. The coding techniques for procedural control can be further classified into the following two categories.

Explicit control describes the execution order of rules explicitly. The common method is to embed data dependencies into rules. A special WME is introduced so that one rule creates the WME and another rule refers to it. Since explicit control requires additional conditions and actions, the rules become more complex. Furthermore, since related descriptions are embedded into different rules, rules that seem independent at first glance may tightly depend on each other. These additional interrelationships decrease the readability of production system programs.

Implicit control realizes procedural control through conflict resolution strategies. Note that conflict resolution was originally introduced to describe the search strategies. However, since a large part of application programs do not require a state space search, conflict resolution is often used for implicit procedural control. In this technique, though rules can retain their simplicity, the control structure hidden behind rules becomes very difficult to understand.

```
(p  pre-selection
    (task ^name select-goods)
      -->
    (make counter ^value 0))

(p  selection-heuristics
    (task  ^name select-goods)
    {<count> (counter ^value {<last> < 100})}
         ;selection heuristics are written here
      -->
         ;register selected goods
    (modify <count> ^value (compute <last> + 1)))

(p  post-selection
    (task ^name select-goods)
    {<count> (counter)}
      -->
    (remove <count>))
```

Figure 5.1: Example Coding Techniques

5.2.2   Example Coding Techniques

Figure 5.1 shows the OPS5 [Forgy, 1981] rules needed to select 100 goods from a number of candidates. This example is not exceedingly complex, and employs rather standard coding techniques already described in text-books [Brownston *et al.*, 1985]. However, since explicit/implicit controls are involved, one can imagine the difficulty of maintaining programs when the techniques are frequently applied.

- Explicit control can be seen in `pre-selection`, which should be the first of the three rules to be fired. This firing order is guaranteed by the special WME `counter`, which is created by `pre-selection` and referenced by the other two rules, i.e., data dependency relationships are embedded into the rules. The `counter` is also used to control the loop in `selection-heuristics` in which the RHS increments the `counter`, while the LHS checks it as the exit condition of the loop.

- Implicit control can be seen in `post-selection`. This rule should not be fired until the 100 goods have been selected. To guarantee this, the `post-selection` rule is written so that it is to be the last rule selected, i.e., the LHS of `post-selection` subsumes the LHS of `selection-heuristics`. Since the OPS5 conflict resolution strategy gives higher priority to more specific rules, `post-selection` is not fired until all goods have been selected.

5.3 Procedural Control Macros

To reduce the difficulty created by the conventional coding techniques, recent expert system building tools provide various control facilities for production systems. These facilities explicitly specify the priorities of rules, transfer the control between sets of rules, called *rulesets*, or enable users to define their own conflict resolution strategies. However, unguided utilization of these new control facilities often decreases the readability of production systems. A more direct approach should be taken to solve the control problem, that is, to separate the description of procedural control from declarative production rules. However, explicit communication facilities have not been provided between control plans and production rules. This is because, unlike ordinary subroutine calls, since rule firings depend on working memory changes, control plans cannot determine in advance which production rule will actually be fired. Thus, the communication facilities that enable control plans to adapt the run-time behavior of production systems are required.

The meta-level control of production systems has been realized by introducing various *procedural control macros (PCMs)* into procedural programming languages: the *rule invocation macro* for executing a single rule; the *rule selection macro* for selecting one executable rule from multiple candidates; the *concurrent execution macro* for creating concurrent activities in a single production system agent. and the *plan definition macro* for developing large scale rule-based systems. Though these macros can be embedded into any procedural language, Lisp is used as the base language in the following examples.

5.3.1   Rule Invocation Macro

?-Macro

Since rules can be viewed as independent processes, the *?-macro*, which is based on the CSP *input command*, is introduced to invoke a single rule.

(? *rule-name* [*variable-binding* ...])

This macro executes the specified rule, and returns *t* when the conditions of the rule are satisfied. Otherwise the *?-macro* monitors the WM and waits for data changes until the conditions of the rule are satisfied. Thus, the *?-macro* is capable of firing rules in dynamically changing environments. For example, data received from sensors in real-time expert systems or from students in interactive tutoring systems are easily handled by *?-macros*.

$-Macro

The *$-macro* is introduced to abandon rule invocation when the conditions of the specified rule are not satisfied. In this case, the *$-macro* returns *nil*. Its syntax is the same as that of the *?-macro*.

($ *rule-name* [*variable-binding* ...])

Conventional coding techniques for procedural control can be easily represented by using *$-macros* in procedural procedural languages. The following two examples written in Lisp represent sequential and conditional rule executions (by utilizing the **progn** special form and the **cond** macro of Lisp), respectively.

(**progn** *$-macro* ... *$-macro*)
(**cond** (*$-macro form*) ... (*$-macro form*))

Since *$-macros* allow users to utilize the original control facilities of procedural languages, various coding tricks to embed control information into rules are no longer required. For example, the program shown in Figure 5.1 can be dramatically simplified: the **pre/post-selection** rules and the **counter** operations in **selection-heuristics** are replaced by the following control plan (which is written using the **dotimes** macro of Lisp).

Plan-1: (dotimes (i 100) ($ selection-heuristics))

Communication between Plans and Rules

Communication between rule invocation macros and production rules is realized by specifying the *variable-bindings*. Variables in rules are called *pattern variables*, while in plans they are *plan variables*. Two kinds of communication are possible as follows:

From plans to rules: Control plans can restrict variable bindings in the rule to be fired. For example, the following *?-macro* filters instantiations by restricting the value of the pattern variable last to the value of *form*.

> (? selection-heuristics :last *form*)

As a result, only rule instantiations with the variable binding of last=*form* are allowed to be fired.

From rules to plans: Rules can bind plan variables to the values of pattern variables. For example, in the following case, after executing selection-heuristics, the value of the pattern variable last is transferred to the *plan-variable*.

> (let ((*plan-variable* *U*))
> (? selection-heuristics :last *plan-variable*))

This data transfer is performed only when the *plan-variable* is unbound. The unbound state is indicated by the special plan variable *U*.

These communication facilities can keep rules independent from control plans. During the communication between plans and rules, rules do not depend on plans, but plans filter rule instantiations or access values held in the rules.

5.3.2   Rule Selection Macro

select-Macro

The production system interpreter tests the LHSs of multiple rules simultaneously and selects one executable rule through conflict resolution. This fundamental mechanism cannot be expressed by any combination of the rule invocation macros and the original control facilities of procedural programming languages. For viewing control plans as a natural extension of conventional production system interpreters, the *rule selection macro (select-macro)* is created, which is influenced by the CSP *guarded command* and *alternative command*.

```
(select (rule-invocation-macro [form])
            ⋮
        (rule-invocation-macro [form])
        [(otherwise [form])])
```

The *select-macro* selects one executable rule from the specified *rule-invocation macros* as follows.

- When several rules become executable, the *select-macro* requests the production system interpreter to choose one rule through conflict resolution. If some rule is invoked, then it returns the value of the subsequent *form*.

- However, if there is no executable rule and if the *select-macro* only contains $-*macros*, it returns *nil*. On the other hand, if there is no executable rule and if the *select-macro* contains some ?-*macros*, it waits for data changes.

- If there is no executable rule and if the **otherwise** clause is specified, then even though ?-*macros* are contained, the *select-macro* evaluates the **otherwise** clause, and returns the value of the *form*. Therefore, when the otherwise clause is specified, ?-*macros* in the *select-macro* behave exactly the same as $-*macros*.

The rule selection macro is simple, but it can represent the behavior of production system interpreters. For example, the interpreter for the rules in Figure 5.1 can be expressed as follows:

Plan-2: (loop (select ((? pre-selection))
                       ((? selection-heuristics))
                       ((? post-selection))
                       (otherwise (return))))

This control plan repeatedly executes the three rules until no rule can be fired. Since *select-macros* can appear at any place in the control plans, and any Lisp *forms* can appear in *select-macros*, conventional production system interpreters can be easily extended and invoked them from anywhere in the control plans.

5.3.3 Concurrent Execution Macro

The following two macros are introduced to create concurrent activities in a single production system agents.

activate-Macro

The **activate** macro is introduced to represent *the receptionist-like tasks*: different processes are triggered by different data arrivals. The syntax is the same as **select**.

    (activate (*rule-invocation-macro* [*form*])
                   :
              (*rule-invocation-macro* [*form*])
              [(otherwise [*form*])])

When one of the executable rules is performed, a process is created to evaluate the subsequent *form*. By defining a rule to receive messages, and by specifying the **select** or **activate** macro in a loop, a concurrent object oriented model can be simulated.

fork-Macro

The `fork` macro is introduced to divide a large task to small pieces. The syntax is as follows.

   (`form` *form*

         :

      form)

   The `fork` macro concurrently executes multiple *forms* in different processes.

5.3.4   Plan Definition Macro

defplan-Macro

When the rule base grows, the control plan becomes larger. The *defplan-macro* is introduced to decompose a large control plan into several modules.

   (`defplan` *plan-name* ()

         (`use` *ruleset-name* ...) *form* ...)

   Each *defplan-macro* defines *a plan module*. The rule invocation / selection macros can appear only within the *defplan-macro*. The related *rulesets* should be declared at the beginning of the *defplan-macro*. The control plans defined by the *defplan-macros* work like Lisp functions; control plans and Lisp functions can invoke each other without any restriction.

   In conventional tools, *rulesets* are defined as units for executing rules as well as for storing rules. As a result of introducing *defplan*, the two roles can be clearly separated: *ruleset* now represents a unit for storing rules and *defplan* defines a control plan unit. In this chapter, therefore the *ruleset* can contain the rules that will not to be executed. This new feature enables users to incrementally debug a large number of rules. Furthermore, large scale shared rule bases become feasible: users can select the rules needed for each application from the shared rule bases by describing the control plans. Since the shared rule bases only contain declarative knowledge, the application programs (i.e., control plans) can be developed and maintained independently from the rule bases.

5.4 Meta-level Control Mechanism

Meta-level control is achieved through combining the procedural programming language, which runs control plans, and the production system interpreter, which fires rules. Note that the WM belongs to the production system interpreter, and thus control plans cannot access the WM directly. This restriction is introduced to clearly distinguish the roles of plans and rules. Figure 5.2 represents the meta-level control mechanism, where the procedural programming language (Lisp) and production system interpreters communicate with each other and cooperatively execute plans and rules. However, this does not mean the meta-level control architecture always requires a multi-process environment. If the concurrent execution macros are not utilized, the simple implementation can be applied the procedural programming language interpreter call the production system interpreter with a PCM each time to fire a rule. Appendix A describes the experience of applying the simple implementation to maintain a single agent production system. The result shows that the meta-level control is also effective to improve the maintainability of rule base systems.

The following procedure determines how plans and rules are executed. The numbered steps correspond to the numbers in Figure 5.2.

1. The procedural programming language interpreter first evaluates a control plan.

   (a) When evaluating `fork`, the procedural programming language interpreter creates the necessary number of child interpreter processes.

   (b) When evaluating rule invocation functions, such as in `select` and `activate`, the procedural programming language interpreter requests the production system interpreter to execute desired rules. In Figure 5.2, rules a to f are requested.

2. The production system interpreter continuously repeats the *Match-Select-Act cycle*.

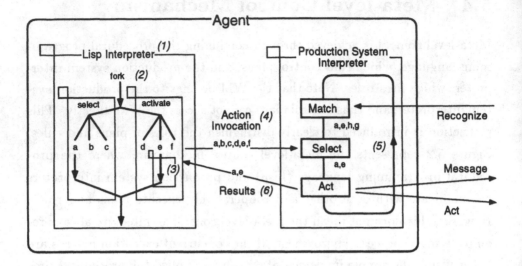

Figure 5.2: Meta-Level Control Mechanism

(a) If there is no executable rule in the requested list, the production system interpreter waits for database changes that satisfy the conditions of the requested rules.

(b) On the other hand, if there is more than one executable rules, the rule requested by the highest priority plan is chosen for execution.

3. When the production system interpreter executes some rule, the result is reported to the requesting procedural programming language interpreter process. In Figure 5.2, rules **a** and **e** are executed.

   (a) When some rule in `activate` is executed, the procedural programming language interpreter creates a child interpreter process to evaluate the subsequent *form*.

   (b) Otherwise, the procedural programming language interpreter continues the execution of the subsequent plan.

In the above mechanism, the procedural programming language interpreter interrupts the production system interpreter at each production cycle, but the overhead is not significant. This is because the RETE network is continuously preserved by the production system interpreter. As a result, there is no overhead in the Match phase, which consumes up to 90% of the total execution time. However, relatively small overheads exist in the Select phase and in the communication between the procedural programming language and production system interpreters. On the other hand, since rules have been simplified, the number of production cycles or at least the cost of the Match phase is reduced. As a result, for example, when applying *Plan-2* in Section 5.3.2, the execution time increases by 20%. However, when applying *Plan-1* in Section 5.3.1, since rules in Figure 5.1 are significantly simplified, the total execution time is reduced by 10%.

5.5 Network Operation Domain

5.5.1 Multiagent Operation

At present in NTT, a Japanese telecommunication agency, various *domain specific operation systems* are being developed. The domains include switching, transmission, traffic control, etc. Expert system technologies have also been introduced into fault diagnosis areas.

Figure 5.3(a) illustrates the current status of network operation centers. Some centralized operation centers are each responsible for the operations of several unmanned offices scattered over wide areas. Human operators are being freed from specific operations, because of the development of domain specific systems. However, they are now being forced to handle more complex *supervising operation tasks*: to accept multiple fault reports asynchronously occurring at different points, to handle various domain specific operation systems, to communicate with neighboring operation centers, and to solve problems from a global point of view. The goal of the rest of this chapter is to develop a multiagent system that manages the supervising network operation tasks currently performed by human operators.

GTE researchers reported a traffic control problem in circuit switched networks, and discussed conflict resolution among distributed agents [Adler *et al.*, 1988]. In this chapter, by contrast, supervising operation tasks are discussed that integrate various domain specific operations. Figure 5.3(b)

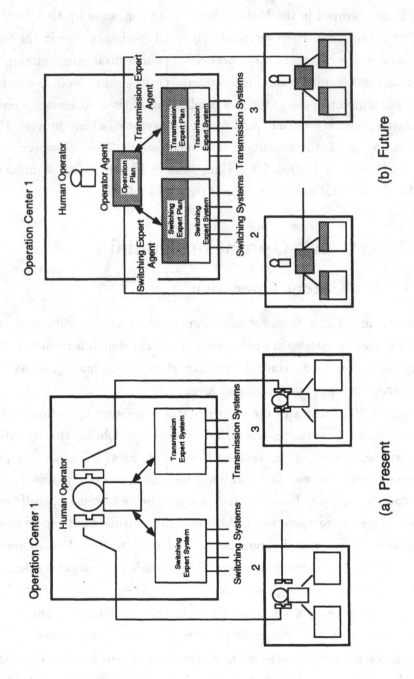

Figure 5.3: Network Operation Center

illustrates future operation centers. Two kinds of agents will be implemented: *operator agents* to act for human operators, and *expert agents* to control domain specific operation systems. Telephone calls among human operators, and interactions with domain specific operation systems will be replaced by communications among multiple automated agents. The role of human operators will be reduced to monitoring and assisting the multiagent operation system.

Figure 5.4 gives an example of multiagent plans for network operations. This figure originally appeared in the *"Electronic switching system D70 maintenance manual"* generated by the NTT Hokkaido Office. Details are not important, but the figure shows how operators in a switching section and in a transmission section repair a fault cooperatively. The directed edges between the two sides of the figure indicate telephone calls between operators. The operation plans are multiagent plans, which cannot be completed by a single agent, but require tight cooperation with other agents. In the case of network operations, since the flow of tasks has been closely analyzed on paper, tightly cooperative multiagent plans can be created. The following is an example of cooperative network operations, which will be used in the rest of this chapter.

Operation Task Example:

Suppose a fault occurs in a common signal link between two operation centers. Fault messages appear at both centers. Operators in both centers start multiagent plans independently. The cause of the two messages is the same, but initially the operators are not aware of this fact. The goal of each plan is to determine the cause of the fault message, and to fix it. The fault may be in the transmission system or in the switching system of either operation center. Thus, each operator concurrently sends messages to related professionals in both centers to ask them to diagnose their facilities.

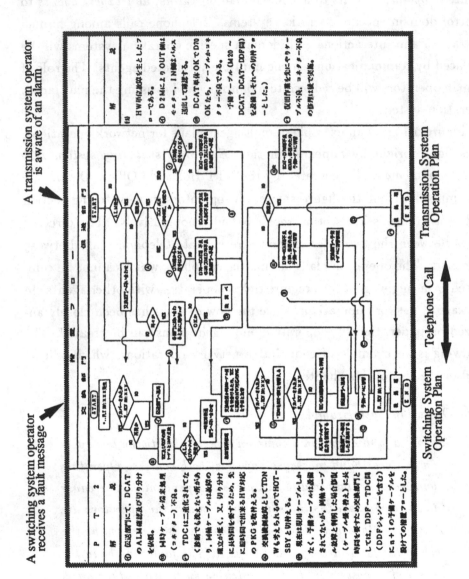

A switching system operator
receives a fault message

A transmission system operator
is aware of an alarm

Switching System   Telephone Call   Transmission System
Operation Plan   Operation Plan

Figure 5.4: Multiagent Network Maintenance

5.5.2 Problem Characteristics

The following characteristics can be observed in the above operation task example.

Incomplete plan:

> The status of networks always changes with new events and concurrently executed multiagent plans. It is thus impossible to define operation plans so completely that they will accept any network state. Invoked plans have no guarantee of being completed. Unexpected events might occur during plan execution so that the requested action may not be performed. In this case, operators must either wait for the situation to change so that the requested action can be restarted or modify their plans to handle the new network status.

Concurrent plan execution:

> In a distributed environment, multiagent plans are asynchronously invoked and concurrently executed. In the example above, the operator divides his/her task among multiple processes and asks other agents to concurrently diagnose related equipment. This shows that more than one process can be created to fix a single fault. Furthermore, at network operation centers, multiple fault repairing processes can be concurrently invoked. A centralized operation center, which manages a wide network area, has more chance of encountering multiple faults. Since the operation plans include human operations, such as package changing, some plans may take a half day to complete. Thus, operators cannot perform the operation plans one by one, but are required to execute them concurrently to minimize down time.

Cooperative meta-level control:

> In the above example, one fault creates two messages, each of which independently invokes a multiagent plan at each operation center. In a more serious case, a single transmission fault will force connected equipment to display a number of messages and alarms; this may panic the operators. Since multiple plans are invoked by the same

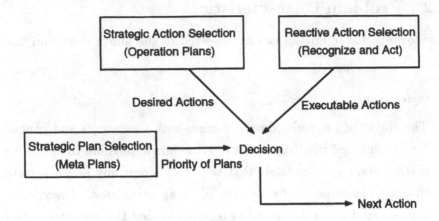

Figure 5.5: Action Model of CoCo Agents

fault, those plans will, most likely, conflict with or reinforce each other. Thus, meta-level control is required to cooperatively focus the attention of multiple agents on the most effective plans.

5.6   Network Operation Agents

This section attempts to solve the problems listed in Section 5.5. An experimental multiagent system, called *CoCo*, is presented. CoCo is a tool for describing, executing and monitoring cooperative operation plans, and is also designed as a research testbed for further cooperative meta-level control problems. CoCo is currently running under the Genera multi-process environment on a Symbolics workstation.

5.6.1   Action Model

Figure 5.5 illustrates the action model of the CoCo agents. The key ideas are as follows.

- A single agent selects an appropriate action for execution by concurrently performing three activities: *reactive action selection, strategic action selection* and *strategic plan selection*. The action is the result of the integration of these three different activities.

- *Strategic action selection* proposes actions desired by operation plans. On the other hand, *reactive action selection* detects executable actions by reactively recognizing the current real world. Actions are performed if they are both desired and executable. In other words, the strategic and reactive activities of an agent are restricted by each other.

- The aim of *strategic plan selection* is to cooperatively focus the multiple agents' attention on a particular operation plan. *Meta-plans* prioritize *operation plans* through communications with other agents.

5.6.2 Agent Configuration

Figure 5.6 displays the configuration of the CoCo agent. There is no memory shared among agents. Agents communicate through messages. CoCo helps to group agents (a group of agents is called *a community*) to restrict the range of broadcasts. Each agent consists of the following components:

- *A database* that represents the agent's view of the real world. Each agent has its own database. The status of the real world is reported to the agent through sensors or other data input facilities. Received messages are also stored in the database. Agents update their views in real time, and examine the stored data to recognize the real world. In Chapter 4, *concurrency control mechanisms* are introduced to maintain consistency of multiagent production systems. This approach allows multiple agents to share the same WM information. Inconsistencies may also exist among agents' views and the real world, but this chapter does not take the concurrency control approach. This is because, in the case of real world sensing, *locks* can only delay the recognition of changes in the real world.

- *An action interpreter* that performs primitive actions defined as production rules. The rule description prevents actions from running away in a dynamic environment: Actions are only executed when preconditions of rules are satisfied.

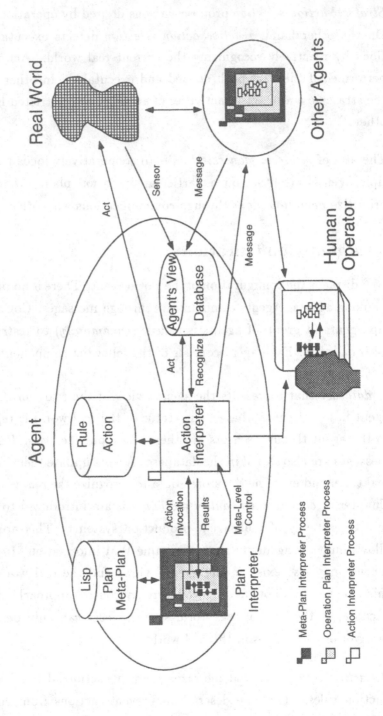

Figure 5.6: Configuration of CoCo Agents

- *A plan interpreter* that evaluates meta-plans and operation plans, each of which is responsible for an appropriate part of a multiagent operation task. Each meta-plan selects an appropriate operation plan, which guides the execution of actions. Several programming facilities are added to Common Lisp to describe the concurrency in the meta-plans or the operation plans.

Each agent contains one action interpreter process, but can contain multiple plan interpreter processes. In other words, all action invocation requests from multiple plan interpreter processes are performed by a single production system. This architecture enables the CoCo agent to efficiently recognize the real world, and to execute a large number of concurrent plans.

For implementing CoCo, a new programming language is not introduced, but two well-proved languages are combined: Lisp and production systems. As a result, CoCo has little problem with language facilities or performance. Furthermore, since CoCo is an extension of Lisp and production systems, it is flexible for future extensions. This is an important feature for any research testbed.

5.7 Experiments

5.7.1 Multiagent Rule Firings

This section represents the CoCo plan description language, which aims to describe both the meta-plans and the operation plans. Figure 5.7 represents an operation plan of the example multiagent operation plan, which is given in Figure 5.4.

Actions

A primitive action is defined by a production rule: *the left hand side (LHS)* represents conditions of an action, and *the right hand side (RHS)* represents how the real world and the database are to be changed.

1. In CoCo, the action description consists of the following three parts. `:trigger` represents *data driven conditions* as in *OPS5* [Forgy, 1981].

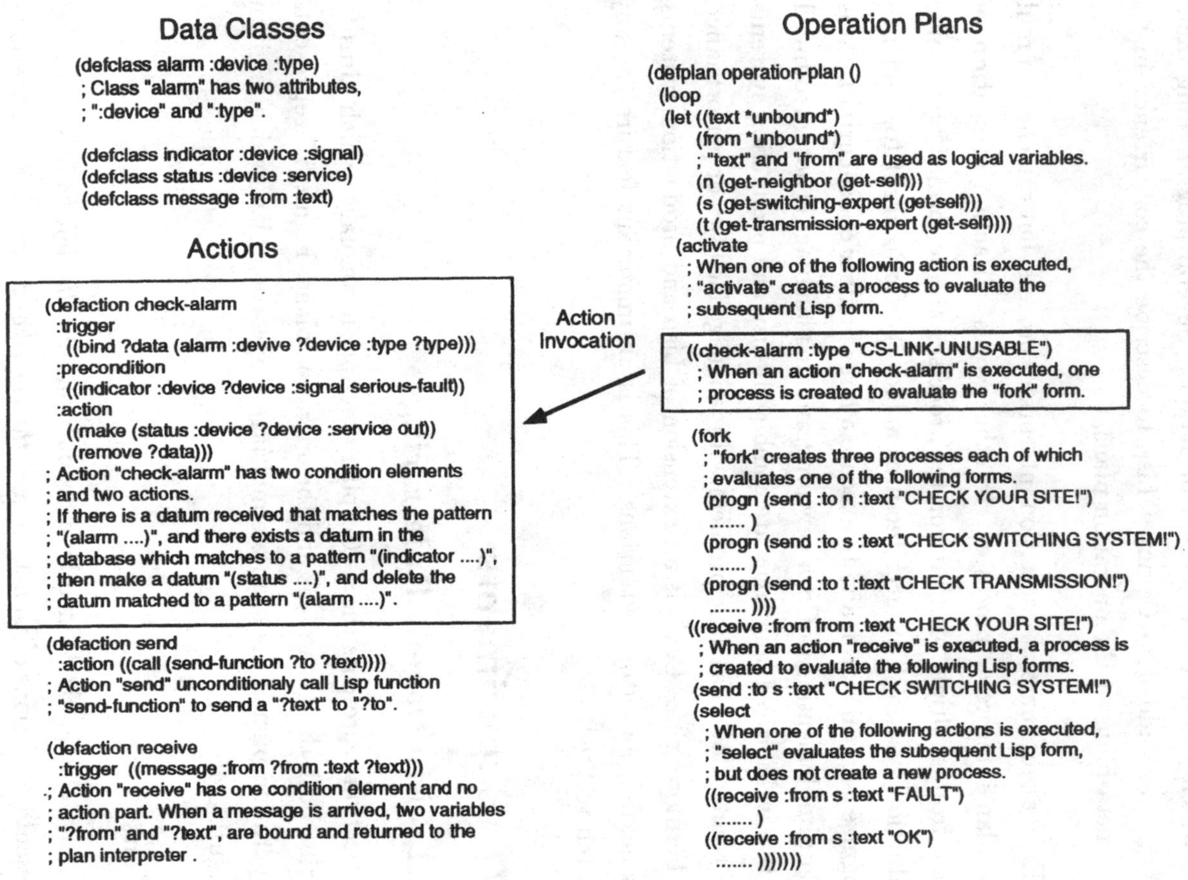

Data Classes

```
(defclass alarm :device :type)
; Class "alarm" has two attributes,
; ":device" and ":type".

(defclass indicator :device :signal)
(defclass status :device :service)
(defclass message :from :text)
```

Actions

```
(defaction check-alarm
:trigger
  ((bind ?data (alarm :device ?device :type ?type)))
:precondition
  ((indicator :device ?device :signal serious-fault))
:action
  ((make (status :device ?device :service out))
   (remove ?data)))
; Action "check-alarm" has two condition elements
; and two actions.
; If there is a datum received that matches the pattern
; "(alarm ....)", and there exists a datum in the
; database which matches to a pattern "(indicator ....)",
; then make a datum "(status ....)", and delete the
; datum matched to a pattern "(alarm ....)".

(defaction send
:action ((call (send-function ?to ?text))))
; Action "send" unconditionally call Lisp function
; "send-function" to send a "?text" to "?to".

(defaction receive
:trigger ((message :from ?from :text ?text)))
; Action "receive" has one condition element and no
; action part. When a message is arrived, two variables
; "?from" and "?text", are bound and returned to the
; plan interpreter.
```

Action Invocation

Operation Plans

```
(defplan operation-plan ()
(loop
  (let ((text *unbound*)
        (from *unbound*)
; "text" and "from" are used as logical variables.
        (n (get-neighbor (get-self)))
        (s (get-switching-expert (get-self)))
        (t (get-transmission-expert (get-self))))
  (activate
; When one of the following action is executed,
; "activate" creats a process to evaluate the
; subsequent Lisp form.

   ((check-alarm :type "CS-LINK-UNUSABLE")
; When an action "check-alarm" is executed, one
; process is created to evaluate the "fork" form.

    (fork
; "fork" creates three processes each of which
; evaluates one of the following forms.
     (progn (send :to n :text "CHECK YOUR SITE!")
      ..... )
     (progn (send :to s :text "CHECK SWITCHING SYSTEM!")
      ..... )
     (progn (send :to t :text "CHECK TRANSMISSION!")
      ..... ))))

   ((receive :from from :text "CHECK YOUR SITE!")
; When an action "receive" is executed, a process is
; created to evaluate the following Lisp forms.
    (send :to s :text "CHECK SWITCHING SYSTEM!")
    (select
; When one of the following actions is executed,
; "select" evaluates the subsequent Lisp form,
; but does not create a new process.
     ((receive :from s :text "FAULT")
      ..... )
     ((receive :from s :text "OK")
      ..... )))))
```

Figure 5.7: Example Multiagent Operation Plan

:precondition represents *demand driven conditions*, which examine existing data in a database. :action represents a list of data to be added or deleted by the action.

2. Let us consider one action example given in Figure 5.7. The action named check-alarm means, *"if a device signals an alarm, and an indicator shows the device suffers a serious fault, then take the device out of service, and reset the alarm."* In this case, since the status of the device cannot be changed by other agents, the database as well as the real world is updated within the action.

Plans

Actions are invoked from operation plans as follows.

1. To execute an action, *an action invocation function* is specified in a plan. As shown in Figure 5.7, suppose the plan interpreter evaluates an action invocation function, (check-alarm :type "cs-link-unusable"). Then the plan interpreter allows the action interpreter to execute check-alarm, if all conditions are satisfied under the restriction of ?type = "cs-link-unusable".

2. If a new datum, (alarm :device SGU :type "cs-link- unusable"), arrives, and other preconditions are satisfied by the database, the action check-alarm is invoked by the action interpreter, (status :device SGU :service out) is added to the database, and the datum indicating the alarm is removed. On the other hand, if some condition is not satisfied, the action, check-alarm, waits for data changes until all conditions are satisfied by the database.

5.7.2 Monitoring Multiagent Rule Firings

Many research activities have been reported on re-planning in dynamic environments. However, plans as shown in Figures 5.4 and 5.7, are too complex to re-plan, even using state-of-the-art techniques. A more plausible approach is that human operators interact with CoCo agents to detect flawed plans and repair them. Thus, *interfacing human and multiagent systems* is another significant research problem.

The problem here is that automated agents and human operators have totally different points of view. Automated agents can report how multiagent plan fragments are scheduled and executed. However, human operators want to know the overall progress of each multiagent plan. To overcome this problem, the CoCo agents are equipped with the following *monitoring facilities*:

The intra-agent monitor, which is periodically called by the action interpreter, enables agents to recognize flawed plans. First, the monitor detects actions that have been waiting for more than a pre-defined time. Second, the monitor examines *the RETE network* [Forgy, 1981] of the waiting actions, then explains why the actions cannot be executed by reporting what data should exist or be deleted.

The inter-agent monitor enables each agent to recognize what goal its process is striving for. The mechanism is as follows:

- At any moment, processes can set values to their *process attributes*. When child processes are created in `fork` or in `activate`, they inherit process attributes from their parent process.
- When a plan interpreter process requests the action interpreter to create data/messages, its process attributes are attached to those data/messages.
- When the data/messages trigger an action, the process attributes are further inherited by the process that requested the execution of the action.

Suppose a process is aware of an alarm, and sets a *Fault-ID* to its process attribute. Then, related processes can find out, by referring to their process attributes, what fault they are originally working for.

In the current CoCo implementation, since a graphic interface has not been supported, human operators communicate with multiagent systems through text messages. For example, the intra-agent monitor may output the following message.

> Switching-expert-agent at Tokyo Operation Center
> cannot execute the action, change-package,
> for more than 30 minutes. The reason is that
> the datum, (new-package :name SGU), has not been
> found in the database.

Then, the inter-agent monitor adds comments by examining the attributes of the plan interpreter process that requested action execution:

> This action is intended to repair the fault,
> F1034:cs-link-unusable, which was originally
> detected by Yokosuka Operation Center.

5.8 Related Work

There has been little research reported on procedural control of production systems. Zisman [1987] proposed to control production rules with the *Petri Net*. Stolfo [1979] represented control plans with the *State Transition Diagram*, while Georgeff [1982] extended it to the *Recursive Transition Network (RTN)*. The techniques proposed, however, share the same drawbacks. All previous research efforts have attempted to introduce new *control plan description languages*. However, because of the difficulty of introducing a new language and the limited programming ability of the proposed languages, the previous approaches have not been widely accepted by expert system builders.

Available expert system building tools often provide a facility to bridge production systems and procedural programming languages. For example, OPS5 [Forgy, 1981] and Bliss , OPS83 [Forgy, 1985] and C , ART [Clayton, 1987] and Lisp , ECliPS [Soo and Schor, 1990] and Lisp, and YES/L1 [Cruise *et al.*, 1987] and PL1 have been tightly connected. In these systems, procedural programs have a limited ability to call production systems (even though production systems can call procedural programs in their RHSs):

procedural programs are only allowed to start whole production systems. On the other hand, the procedural control macros (PCMs) proposed in this chapter can interrupt production systems at each execution cycle and specify the next rule to be executed.

Many papers have been published on the subject of problem solving in a distributed environment. *DVMT* [Lesser *et al.*, 1983] and *MACE* [Gasser *et al.*,1987] are distributed artificial intelligence testbeds which test a wide range of coordination and organizational techniques. *PRS* [Georgeff and Lansky, 1986] and *MI Architecture* (multi-processor interrupt driven architecture) [Hayes-Roth *et al.*, 1989] are adaptive problem solving systems, which are capable of handling dynamically changing situations. Multiagent approaches, like that addressed in this chapter, have also attracted the attention of researchers working on so-called *feature interaction* or *service interaction* problems [Bowen *et al.*, 1989; Griffeth, 1991], where multiple independently specified services interact with each other. At present, however, no existing system could have provided enough facilities for the cooperative operation tasks addressed in Section 5.5.2.

5.9   Summary

A meta-level control architecture has been proposed, which is superior to previous solutions, such as using the *State Transition Diagrams* or the *Petri Nets*, in two ways:

- *The procedural control facilities of procedural programming languages can be utilized.* Since the control facilities have been continually enhanced, utilizing them is advantageous both for implementing a new control plan description language and having it widely accepted.

- *The Procedural Control Macros are easier to implement than the State Transition Diagrams or the Petri Nets.* By adding small code segments to the existing procedural programming language and production system interpreters, it is shown that powerful meta-level control mechanisms can be obtained.

Every day activities in telecommunication network operation centers have been analyzed, and a model for cooperative multiagent operations has been proposed. Since no system exists that can describe and execute network operation tasks, an experimental multiagent system, called CoCo, has been implemented. The next step is to extend CoCo to heterogeneous network environments.

Appendix A

Meta-Level Control for Rulebase Maintenance

A.1　The CAIRNEY Rulebase

After a decade of research on production systems, a number of expert systems are now in their maintenance phases. While the performance improvement of production systems has widely been studied [Forgy, 1982; Miranker, 1987; Ishida 1988], the readability and maintainability of application programs have not yet been thoroughly investigated. As a result, recent production system applications have been experiencing exceedingly difficult software maintenance problems. Soloway *et al.* [1987] have reported that 50% of XCON's rules are updated every year. One of the major reported difficulties in maintaining such a large-scale expert system is that the execution control of rule firings, either implicit or explicit, is buried in the rules themselves. The same problem as XCON's was faced during the development of the intelligent tutoring system CAIRNEY [Fukuhara *et al.*, 1990].

CAIRNEY is a computer-aided tutoring system that aims to provide private lessons appropriate to each student. Most of the commercial tutoring systems available display variable teaching materials using the *branching method*, where the ordering of materials is precisely planned before execution. The shortcoming of this method is that a substantial volume of tutoring plans, involving a sufficient number of branches, is required for implementing private lessons. CAIRNEY employs the *rule-based method*, in which tutoring plans are written in production rules that adaptively select teaching materials based on the students' levels of understanding and learning histories.

The CAIRNEY project started in 1984. After refining the rulebase for two years, a prototype system called MASTERS was completed in 1986. The effectiveness of this system was confirmed by 32 students in 1987 [Morihara *et al.*, 1987]. The project matured and the system was renamed CAIRNEY when it was put into practical use. CAIRNEY is currently in operation at several tens of NTT offices, and is planned to be introduced to more than one hundred offices in a few years. Around eighty professionals are currently developing teaching materials that mainly address public telecommunication network operations. To date, more than forty courses have been completed.

The CAIRNEY rulebase has been rebuilt twice using different production systems over the last ten years. Since various coding tricks were embedded into the rules, CAIRNEY started encountering serious maintenance problems. To address these problems, the CAIRNEY rulebase had to be made easy-to-enhance. Thus, a *meta-level control architecture* for production systems (see Chapter 5) was applied. The rulebase was redescribed by utilizing the proposed meta-level control architecture. The results show that the readability and maintainability of the rulebase were increased by separating control plans and declarative rules.

A.2   Representation and Run-Time Overheads

The CAIRNEY rulebase currently contains approximately 500 rules. To enhance the maintainability of the rulebase, the 370 rules that were frequently executed were redescribed. This effort aimed at separating the CAIRNEY rulebase into procedural control plans and declarative domain rules. The file volume and execution speed of the rules were measured both before and after the redescription. This was done to determine the representation and run-time overheads of the meta-level control architecture. The major results obtained from this experiment are summarized as follows.

Representation Overheads:

After the redescription of the 370 rules, the file volume decreased from $100Kbytes$ to $74Kbytes$, while the newly created control plans occupied $20Kbytes$. Since the total volume of the CAIRNEY rulebase slightly decreased, *no significant representation overhead is introduced by the meta-level control architecture.*

Run-time Overheads:

The run-time overhead has been evaluated along the most frequently executed route of the CAIRNEY rulebase. The route selects the teaching material to be presented next. Before the redescription, the route contained an average of 21 rule firings. After the redescription, procedural control macros (PCMs) are called 20 times, and the number of the rule firings was reduced to 9: Eleven PCM calls are performed to confirm that the specified rules are not to be fired. The reduction in the number of rule firings almost offsets the overall overhead of the meta-level control architecture. In total, the redescribed rulebase can select the next teaching material with less than 3% increase in overhead. Though the overhead depends on the application, this experiment supports our expectation: *no significant run-time overhead is introduced by the meta-level control architecture.*

A.3 Readability and Maintainability Enhancements

Based on the redescription experiment, the readability and maintainability enhancements achieved by the meta-level control architecture were determined. The original and redescribed rulebases were precisely investigated and compared. Furthermore, the amount of effort required by a version-up of the CAIRNEY rulebase was measured for both rulebases. The version-up aimed to reflect a dozen user requests. The evaluation results indicated the impact of introducing the meta-level control architecture.

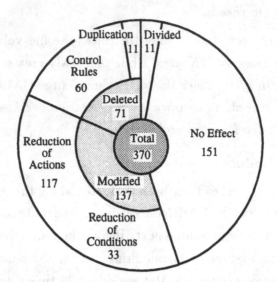

Figure A.1: Changes in the Number of Rules

Readability Enhancement:

Figure A.1 illustrates the effect of the meta-level control architecture on the number of rules. The original 370 rules were reduced to 310 rules: *71 rules were eliminated and 11 rules were created.* Sixty of the 71 rules contained only control information, and thus were replaced by the control plans. This concentration improves the readability of both the control plans and production rules.

It is important to note that 11 more rules were deleted, though they contained declarative knowledge. This is because the redescription revealed that the declarative knowledge in the rules was redundant; since the rules had to be fired in different contexts, different control information was embedded into the rules. Thus, it should be pointed out that *the meta-level control architecture not only concentrates the embedded control information into control plans, but also eliminates the duplication of declarative knowledge that often causes maintenance problems.*

Figure A.2 represents the effect of the redescription on the complexity of each rule. The number of condition/action elements of the remaining rules were reduced by 42%. This fact shows that *the meta-level*

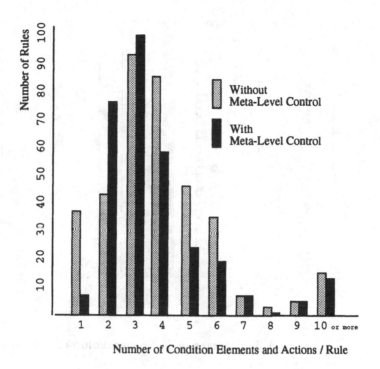

Figure A.2: Changes in Rule Complexity

control architecture not only reduces the number of rules but also significantly simplifies the complexity of each rule.

Maintainability Enhancement:

Figure A.3 shows the volume of rules and plans that were referenced by expert system builders during the version-up of the CAIRNEY rulebase. The reference volume represents the cost of understanding both control and declarative knowledge during the maintenance process. Figure A.3(a) indicates that the total number of referenced rules and plans is reduced by 38%, while Figure A.3(b) shows that the number of referenced lines decreased by 19%. However, the effect of the meta-level control is not limited to these numbers. It is important to remember that the control information was buried throughout the original rulebase. After redescription, however, the control informa-

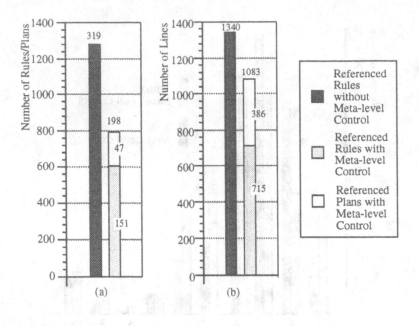

Figure A.3: Changes in Reference Volumes

tion was gathered into the control plans and took up $20Kbytes$. By comparing the version-up process for the original rulebase to that of the control plans, it appears that the reference volume necessary to understand the control flow of the CAIRNEY rulebase is reduced by 71% to 85%. The above measurements confirm that *the meta-level control architecture significantly reduces the effort needed to maintain production systems.*

Since control knowledge becomes concentrated into control plans, production rules are only used to describe declarative knowledge. The readability and maintainability of production systems are thus improved without significant representation or run-time overheads. The control plans can be efficiently compiled by procedural language compilers, and thus the increase in overheads is minimized. Furthermore, it has also been shown that the effect of simplifying rules can offset the total overheads imposed by the meta-level control architecture.

During the ten year experience with production systems, we have heard many complaints about the representation ability of production systems. While getting a new declarative programming paradigm, production system users have been forced to describe procedural knowledge in a declarative representation form. The meta-level control architecture will enable users to fully realize the benefits of the production system paradigm without any significant representation or run-time overheads.

Appendix B

Waltz's Labeling Program

```
;;;;;;;;;;;;;;;;;;;;;;;;;;;;;;;;;;;;;;;;;;;;;;;;;;;;;;;;;;;;;;;;;
;;;;;;;;;;;;;;;;;;;;;;;;;;;;;;;;;;;;;;;;;;;;;;;;;;;;;;;;;;;;;;;;;
;                                                               ;
;   Data & Knowledge Structure for Waltz's Algorithm            ;
;                                                               ;
;;;;;;;;;;;;;;;;;;;;;;;;;;;;;;;;;;;;;;;;;;;;;;;;;;;;;;;;;;;;;;;;;
;;;;;;;;;;;;;;;;;;;;;;;;;;;;;;;;;;;;;;;;;;;;;;;;;;;;;;;;;;;;;;;;;

(literalize possible-junction-label
   junction-type line-1 line-2 line-3)
(literalize junction junction-type
   junction-ID line-ID-1 line-ID-2 line-ID-3)
(literalize labelling-candidate
   junction-ID line-1 line-2 line-3)


;;;;;;;;;;;;;;;;;;;;;;;;;;;;;;;;;;;;;
;  Possible Junction Labeling       ;
;;;;;;;;;;;;;;;;;;;;;;;;;;;;;;;;;;;;;

(p initialize
   (stage initialize)
   -->
   (remove 1)
   (make stage make-data)

;                          \    /
; Junction type : L       1 \  / 2
;                           V

   (make possible-junction-label
      ^junction-type L ^line-1 out ^line-2 in ^line-3 nil)
   (make possible-junction-label
      ^junction-type L ^line-1 in ^line-2 out ^line-3 nil)
   (make possible-junction-label
      ^junction-type L ^line-1 + ^line-2 out ^line-3 nil)
   (make possible-junction-label
      ^junction-type L ^line-1 in ^line-2 + ^line-3 nil)
```

```
    (make possible-junction-label
        ^junction-type L ^line-1 - ^line-2 in ^line-3 nil)
    (make possible-junction-label
        ^junction-type L ^line-1 out ^line-2 - ^line-3 nil)

;                                  1 \ / 3
; Junction type: FORK                V
;                                   2 |

    (make possible-junction-label
        ^junction-type FORK ^line-1 + ^line-2 + ^line-3 +)
    (make possible-junction-label
        ^junction-type FORK ^line-1 - ^line-2 - ^line-3 -)
    (make possible-junction-label
        ^junction-type FORK ^line-1 in ^line-2 - ^line-3 out)
    (make possible-junction-label
        ^junction-type FORK ^line-1 - ^line-2 out ^line-3 in)
    (make possible-junction-label
        ^junction-type FORK ^line-1 out ^line-2 in ^line-3 -)

;                                  1 _____ 3
; Junction type: T                       |
;                                        | 2

    (make possible-junction-label
        ^junction-type T ^line-1 out ^line-2 + ^line-3 in)
    (make possible-junction-label
        ^junction-type T ^line-1 out ^line-2 - ^line-3 in)
    (make possible-junction-label
        ^junction-type T ^line-1 out ^line-2 in ^line-3 in)
    (make possible-junction-label
        ^junction-type T ^line-1 out ^line-2 out ^line-3 in)

;                                     /|\
; Junction type: ARROW            1 / | \ 3
;                                   /  |2 \

    (make possible-junction-label
        ^junction-type ARROW ^line-1 in ^line-2 + ^line-3 out)
    (make possible-junction-label
        ^junction-type ARROW ^line-1 - ^line-2 + ^line-3 -)
    (make possible-junction-label
        ^junction-type ARROW ^line-1 + ^line-2 - ^line-3 +))
```

```
;;;;;;;;;;;;;;;;;;;;;;;;;;;;
; Scene to be Analyzed   ;
;;;;;;;;;;;;;;;;;;;;;;;;;;;;
;
;                  ----------->
;
;                 A              B
;               / \            / \
;           1 /     \ 2    3 /     \ 4
;             /   C    \      /  D   \
;           /  5/|\6    \    /  /|\   \
;        E /10/+|     \  / / | \  \
;          |\+/       \ /  /  7/ 8 | \9 \
;          | | F  |-    \ /  /    |    \
;          | |12 |41 +\ G /+   |-        \
;        11| |+     |     |\ /  M |    +\      \
;          | |14/K\     |     /       \39| O
;        H | |-/    \15  |16 17/    \18  N\+/|
;        /-\|/  P   \-  |+ -/ Q  -   19|  |20  |
;        |  / 13J  /|\       / \          +|  |    |
;      21/  /      22/  | | \ /  \            19|  |20  |
;      /  /      /    | 24\  \|/  \       27     +|  | |    |
;     |  R/    |\30  /+ 23|   +\ T /+ 26 +\      W\|  |    V
;     |  |\+ (/-   -|       | /25 -|        | \ /+  |  |
;     29| \S/     |     U  \ /    V \ /+  |28 |
;       |  31|      /X\   32|+    /Y\   +|  |  |
;       |  |+     |       |     33|  |  |
;     Z \   |  /40     35\  |/36   37\ |/38 / DD
;     34\ | |    |    /    |  /      \ | /
;       \|/        \|/        \|/
;        AA          BB          CC
;
;                  <----------
;

(p make-data
   (stage make-data)
   -->
   (remove 1)
   (make stage enumerate-possible-candidates)

   (make junction ^junction-type L ^junction-ID A
      ^line-ID-1 2 ^line-ID-2 1 ^line-ID-3 NIL)
   (make junction ^junction-type L ^junction-ID B
      ^line-ID-1 4 ^line-ID-2 3 ^line-ID-3 NIL)
   (make junction ^junction-type ARROW ^junction-ID C
      ^line-ID-1 5 ^line-ID-2 41 ^line-ID-3 6)
   (make junction ^junction-type ARROW ^junction-ID D
      ^line-ID-1 7 ^line-ID-2 8 ^line-ID-3 9)
   (make junction ^junction-type ARROW ^junction-ID E
```

```
        ^line-ID-1 11 ^line-ID-2 10 ^line-ID-3 1)
(make junction ^junction-type FORK ^junction-ID F
        ^line-ID-1 10 ^line-ID-2 12 ^line-ID-3 5)
(make junction ^junction-type L ^junction-ID G
        ^line-ID-1 2 ^line-ID-2 3 ^line-ID-3 NIL)
(make junction ^junction-type FORK ^junction-ID H
        ^line-ID-1 11 ^line-ID-2 21 ^line-ID-3 13)
(make junction ^junction-type ARROW ^junction-ID J
        ^line-ID-1 14 ^line-ID-2 12 ^line-ID-3 13)
(make junction ^junction-type FORK ^junction-ID K
        ^line-ID-1 41 ^line-ID-2 14 ^line-ID-3 15)
(make junction ^junction-type FORK ^junction-ID L
        ^line-ID-1 6 ^line-ID-2 16 ^line-ID-3 7)
(make junction ^junction-type FORK ^junction-ID M
        ^line-ID-1 8 ^line-ID-2 17 ^line-ID-3 18)
(make junction ^junction-type FORK ^junction-ID N
        ^line-ID-1 9 ^line-ID-2 19 ^line-ID-3 39)
(make junction ^junction-type ARROW ^junction-ID O
        ^line-ID-1 4 ^line-ID-2 39 ^line-ID-3 20)
(make junction ^junction-type ARROW ^junction-ID P
        ^line-ID-1 2 ^line-ID-2 23 ^line-ID-3 24)
(make junction ^junction-type ARROW ^junction-ID Q
        ^line-ID-1 25 ^line-ID-2 26 ^line-ID-3 27)
(make junction ^junction-type ARROW ^junction-ID R
        ^line-ID-1 29 ^line-ID-2 30 ^line-ID-3 21)
(make junction ^junction-type FORK ^junction-ID S
        ^line-ID-1 30 ^line-ID-2 31 ^line-ID-3 22)
(make junction ^junction-type ARROW ^junction-ID T
        ^line-ID-1 17 ^line-ID-2 16 ^line-ID-3 15)
(make junction ^junction-type FORK ^junction-ID U
        ^line-ID-1 24 ^line-ID-2 32 ^line-ID-3 25)
(make junction ^junction-type FORK ^junction-ID V
        ^line-ID-1 27 ^line-ID-2 33 ^line-ID-3 28)
(make junction ^junction-type ARROW ^junction-ID W
        ^line-ID-1 19 ^line-ID-2 18 ^line-ID-3 28)
(make junction ^junction-type FORK ^junction-ID X
        ^line-ID-1 23 ^line-ID-2 40 ^line-ID-3 35)
(make junction ^junction-type FORK ^junction-ID Y
        ^line-ID-1 26 ^line-ID-2 36 ^line-ID-3 37)
(make junction ^junction-type L ^junction-ID Z
        ^line-ID-1 29 ^line-ID-2 34 ^line-ID-3 NIL)
(make junction ^junction-type ARROW ^junction-ID AA
        ^line-ID-1 40 ^line-ID-2 31 ^line-ID-3 34)
(make junction ^junction-type ARROW ^junction-ID BB
        ^line-ID-1 36 ^line-ID-2 32 ^line-ID-3 35)
(make junction ^junction-type ARROW ^junction-ID CC
        ^line-ID-1 38 ^line-ID-2 33 ^line-ID-3 37)
(make junction ^junction-type L ^junction-ID DD
        ^line-ID-1 38 ^line-ID-2 20 ^line-ID-3 NIL))
```

```
;;;;;;;;;;;;;;;;;;;;;;;;;;;;;;;;;;;;;;;;;;;;;;;;;;;;;;;;;;;;
;;;;;;;;;;;;;;;;;;;;;;;;;;;;;;;;;;;;;;;;;;;;;;;;;;;;;;;;;;;;
;                                                          ;
;        Production Rules for Waltz'slgorithm             ;
;                                                          ;
;;;;;;;;;;;;;;;;;;;;;;;;;;;;;;;;;;;;;;;;;;;;;;;;;;;;;;;;;;;;
;;;;;;;;;;;;;;;;;;;;;;;;;;;;;;;;;;;;;;;;;;;;;;;;;;;;;;;;;;;;

;;;;;;;;;
; Start ;
;;;;;;;;;

(p start-Waltz
   (start)
   -->
   (remove 1)
   (make stage initialize))

;;;;;;;;;;;;;;;;;;;;;;;;;;;;;;;;;;;;;;
; Enumerate Possible Candidates ;
;;;;;;;;;;;;;;;;;;;;;;;;;;;;;;;;;;;;;;

(p enumerate-possible-candidates
   (stage enumerate-possible-candidates)
   (junction ^junction-type <T> ^junction-ID <J>)
   (possible-junction-label ^junction-type <T>
      ^line-1 <L1> ^line-2 <L2> ^line-3 <L3>)
  -(labelling-candidate ^junction-ID <J>
      ^line-1 <L1> ^line-2 <L2> ^line-3 <L3>)
   -->
   (make labelling-candidate ^junction-ID <J>
      ^line-1 <L1> ^line-2 <L2> ^line-3 <L3>))

(p go-to-reduce-candidates
   (stage enumerate-possible-candidates)
   -->
   (remove 1)
   (make stage reduce-candidates))

;;;;;;;;;;;;;;;;;;;;;;;;;
; Reduce Candidates ;
;;;;;;;;;;;;;;;;;;;;;;;;;

(p one-one-plus
   (stage reduce-candidates)
   (junction ^junction-ID <X> ^line-ID-1 <L>)
   (junction ^junction-ID {<Y> <> <X>} ^line-ID-1 <L>)
   (labelling-candidate ^junction-ID <X> ^line-1 +)
  -(labelling-candidate ^junction-ID <Y> ^line-1 +)
   --> (remove 4))

(p one-one-minus
   (stage reduce-candidates)
   (junction ^junction-ID <X> ^line-ID-1 <L>)
   (junction ^junction-ID {<Y> <> <X>} ^line-ID-1 <L>)
```

```
          (labelling-candidate ^junction-ID <X> ^line-1 -)
        -(labelling-candidate ^junction-ID <Y> ^line-1 -)
         --> (remove 4))

(p one-one-in
   (stage reduce-candidates)
   (junction ^junction-ID <X> ^line-ID-1 <L>)
   (junction ^junction-ID {<Y> <> <X>} ^line-ID-1 <L>)
   (labelling-candidate ^junction-ID <X> ^line-1 in)
 -(labelling-candidate ^junction-ID <Y> ^line-1 out)
   --> (remove 4))

(p one-one-out
   (stage reduce-candidates)
   (junction ^junction-ID <X> ^line-ID-1 <L>)
   (junction ^junction-ID {<Y> <> <X>} ^line-ID-1 <L>)
   (labelling-candidate ^junction-ID <X> ^line-1 out)
 -(labelling-candidate ^junction-ID <Y> ^line-1 in)
   --> (remove 4))

(p one-two-plus
   (stage reduce-candidates)
   (junction ^junction-ID <X> ^line-ID-1 <L>)
   (junction ^junction-ID {<Y> <> <X>} ^line-ID-2 <L>)
   (labelling-candidate ^junction-ID <X> ^line-1 +)
 -(labelling-candidate ^junction-ID <Y> ^line-2 +)
   --> (remove 4))

(p one-two-minus
   (stage reduce-candidates)
   (junction ^junction-ID <X> ^line-ID-1 <L>)
   (junction ^junction-ID {<Y> <> <X>} ^line-ID-2 <L>)
   (labelling-candidate ^junction-ID <X> ^line-1 -)
 -(labelling-candidate ^junction-ID <Y> ^line-2 -)
   --> (remove 4))

(p one-two-in
   (stage reduce-candidates)
   (junction ^junction-ID <X> ^line-ID-1 <L>)
   (junction ^junction-ID {<Y> <> <X>} ^line-ID-2 <L>)
   (labelling-candidate ^junction-ID <X> ^line-1 in)
 -(labelling-candidate ^junction-ID <Y> ^line-2 out)
   --> (remove 4))

(p one-two-out
   (stage reduce-candidates)
   (junction ^junction-ID <X> ^line-ID-1 <L>)
   (junction ^junction-ID {<Y> <> <X>} ^line-ID-2 <L>).
   (labelling-candidate ^junction-ID <X> ^line-1 out)
 -(labelling-candidate ^junction-ID <Y> ^line-2 in)
   --> (remove 4))

(p one-three-plus
   (stage reduce-candidates)
   (junction ^junction-ID <X> ^line-ID-1 <L>)
```

```
   (junction ^junction-ID {<Y> <> <X>} ^line-ID-3 <L>)
   (labelling-candidate ^junction-ID <X> ^line-1 +)
 -(labelling-candidate ^junction-ID <Y> ^line-3 +)
   --> (remove 4))

(p one-three-minus
   (stage reduce-candidates)
   (junction ^junction-ID <X> ^line-ID-1 <L>)
   (junction ^junction-ID {<Y> <> <X>} ^line-ID-3 <L>)
   (labelling-candidate ^junction-ID <X> ^line-1 -)
 -(labelling-candidate ^junction-ID <Y> ^line-3 -)
   --> (remove 4))

(p one-three-in
   (stage reduce-candidates)
   (junction ^junction-ID <X> ^line-ID-1 <L>)
   (junction ^junction-ID {<Y> <> <X>} ^line-ID-3 <L>)
   (labelling-candidate ^junction-ID <X> ^line-1 in)
 -(labelling-candidate ^junction-ID <Y> ^line-3 out)
   --> (remove 4))

(p one-three-out
   (stage reduce-candidates)
   (junction ^junction-ID <X> ^line-ID-1 <L>)
   (junction ^junction-ID {<Y> <> <X>} ^line-ID-3 <L>)
   (labelling-candidate ^junction-ID <X> ^line-1 out)
 -(labelling-candidate ^junction-ID <Y> ^line-3 in)
   --> (remove 4))

(p two-one-plus
   (stage reduce-candidates)
   (junction ^junction-ID <X> ^line-ID-2 <L>)
   (junction ^junction-ID {<Y> <> <X>} ^line-ID-1 <L>)
   (labelling-candidate ^junction-ID <X> ^line-2 +)
 -(labelling-candidate ^junction-ID <Y> ^line-1 +)
   --> (remove 4))

(p two-one-minus
   (stage reduce-candidates)
   (junction ^junction-ID <X> ^line-ID-2 <L>)
   (junction ^junction-ID {<Y> <> <X>} ^line-ID-1 <L>)
   (labelling-candidate ^junction-ID <X> ^line-2 -)
 -(labelling-candidate ^junction-ID <Y> ^line-1 -)
   --> (remove 4))

(p two-one-in
   (stage reduce-candidates)
   (junction ^junction-ID <X> ^line-ID-2 <L>)
   (junction ^junction-ID {<Y> <> <X>} ^line-ID-1 <L>)
   (labelling-candidate ^junction-ID <X> ^line-2 in)
 -(labelling-candidate ^junction-ID <Y> ^line-1 out)
   --> (remove 4))

(p two-one-out
   (stage reduce-candidates)
```

```
       (junction ^junction-ID <X> ^line-ID-2 <L>)
       (junction ^junction-ID {<Y> <> <X>} ^line-ID-1 <L>)
       (labelling-candidate ^junction-ID <X> ^line-2 out)
      -(labelling-candidate ^junction-ID <Y> ^line-1 in)
       --> (remove 4))

(p two-two-plus
       (stage reduce-candidates)
       (junction ^junction-ID <X> ^line-ID-2 <L>)
       (junction ^junction-ID {<Y> <> <X>} ^line-ID-2 <L>)
       (labelling-candidate ^junction-ID <X> ^line-2 +)
      -(labelling-candidate ^junction-ID <Y> ^line-2 +)
       --> (remove 4))

(p two-two-minus
       (stage reduce-candidates)
       (junction ^junction-ID <X> ^line-ID-2 <L>)
       (junction ^junction-ID {<Y> <> <X>} ^line-ID-2 <L>)
       (labelling-candidate ^junction-ID <X> ^line-2 -)
      -(labelling-candidate ^junction-ID <Y> ^line-2 -)
       --> (remove 4))

(p two-two-in
       (stage reduce-candidates)
       (junction ^junction-ID <X> ^line-ID-2 <L>)
       (junction ^junction-ID {<Y> <> <X>} ^line-ID-2 <L>)
       (labelling-candidate ^junction-ID <X> ^line-2 in)
      -(labelling-candidate ^junction-ID <Y> ^line-2 out)
       --> (remove 4))

(p two-two-out
       (stage reduce-candidates)
       (junction ^junction-ID <X> ^line-ID-2 <L>)
       (junction ^junction-ID {<Y> <> <X>} ^line-ID-2 <L>)
       (labelling-candidate ^junction-ID <X> ^line-2 out)
      -(labelling-candidate ^junction-ID <Y> ^line-2 in)
       --> (remove 4))

(p two-three-plus
       (stage reduce-candidates)
       (junction ^junction-ID <X> ^line-ID-2 <L>)
       (junction ^junction-ID {<Y> <> <X>} ^line-ID-3 <L>)
       (labelling-candidate ^junction-ID <X> ^line-2 +)
      -(labelling-candidate ^junction-ID <Y> ^line-3 +)
       --> (remove 4))

(p two-three-minus
       (stage reduce-candidates)
       (junction ^junction-ID <X> ^line-ID-2 <L>)
       (junction ^junction-ID {<Y> <> <X>} ^line-ID-3 <L>)
       (labelling-candidate ^junction-ID <X> ^line-2 -)
      -(labelling-candidate ^junction-ID <Y> ^line-3 -)
       --> (remove 4))

(p two-three-in
```

```
(p two-three-in
   (stage reduce-candidates)
   (junction ^junction-ID <X> ^line-ID-2 <L>)
   (junction ^junction-ID {<Y> <> <X>} ^line-ID-3 <L>)
   (labelling-candidate ^junction-ID <X> ^line-2 in)
  -(labelling-candidate ^junction-ID <Y> ^line-3 out)
   --> (remove 4))

(p two-three-out
   (stage reduce-candidates)
   (junction ^junction-ID <X> ^line-ID-2 <L>)
   (junction ^junction-ID {<Y> <> <X>} ^line-ID-3 <L>)
   (labelling-candidate ^junction-ID <X> ^line-2 out)
  -(labelling-candidate ^junction-ID <Y> ^line-3 in)
   --> (remove 4))

(p three-one-plus
   (stage reduce-candidates)
   (junction ^junction-ID <X> ^line-ID-3 <L>)
   (junction ^junction-ID {<Y> <> <X>} ^line-ID-1 <L>)
   (labelling-candidate ^junction-ID <X> ^line-3 +)
  -(labelling-candidate ^junction-ID <Y> ^line-1 +)
   --> (remove 4))

(p three-one-minus
   (stage reduce-candidates)
   (junction ^junction-ID <X> ^line-ID-3 <L>)
   (junction ^junction-ID {<Y> <> <X>} ^line-ID-1 <L>)
   (labelling-candidate ^junction-ID <X> ^line-3 -)
  -(labelling-candidate ^junction-ID <Y> ^line-1 -)
   --> (remove 4))

(p three-one-in
   (stage reduce-candidates)
   (junction ^junction-ID <X> ^line-ID-3 <L>)
   (junction ^junction-ID {<Y> <> <X>} ^line-ID-1 <L>)
   (labelling-candidate ^junction-ID <X> ^line-3 in)
  -(labelling-candidate ^junction-ID <Y> ^line-1 out)
   --> (remove 4))

(p three-one-out
   (stage reduce-candidates)
   (junction ^junction-ID <X> ^line-ID-3 <L>)
   (junction ^junction-ID {<Y> <> <X>} ^line-ID-1 <L>)
   (labelling-candidate ^junction-ID <X> ^line-3 out)
  -(labelling-candidate ^junction-ID <Y> ^line-1 in)
   --> (remove 4))

(p three-two-plus
   (stage reduce-candidates)
   (junction ^junction-ID <X> ^line-ID-3 <L>)
   (junction ^junction-ID {<Y> <> <X>} ^line-ID-2 <L>)
   (labelling-candidate ^junction-ID <X> ^line-3 +)
  -(labelling-candidate ^junction-ID <Y> ^line-2 +)
   --> (remove 4))
```

```
(p three-two-minus
   (stage reduce-candidates)
   (junction ^junction-ID <X> ^line-ID-3 <L>)
   (junction ^junction-ID {<Y> <> <X>} ^line-ID-2 <L>)
   (labelling-candidate ^junction-ID <X> ^line-3 -)
  -(labelling-candidate ^junction-ID <Y> ^line-2 -)
   --> (remove 4))

(p three-two-in
   (stage reduce-candidates)
   (junction ^junction-ID <X> ^line-ID-3 <L>)
   (junction ^junction-ID {<Y> <> <X>} ^line-ID-2 <L>)
   (labelling-candidate ^junction-ID <X> ^line-3 in)
  -(labelling-candidate ^junction-ID <Y> ^line-2 out)
   --> (remove 4))

(p three-two-out
   (stage reduce-candidates)
   (junction ^junction-ID <X> ^line-ID-3 <L>)
   (junction ^junction-ID {<Y> <> <X>} ^line-ID-2 <L>)
   (labelling-candidate ^junction-ID <X> ^line-3 out)
  -(labelling-candidate ^junction-ID <Y> ^line-2 in)
   --> (remove 4))

(p three-three-plus
   (stage reduce-candidates)
   (junction ^junction-ID <X> ^line-ID-3 <L>)
   (junction ^junction-ID {<Y> <> <X>} ^line-ID-3 <L>)
   (labelling-candidate ^junction-ID <X> ^line-3 +)
  -(labelling-candidate ^junction-ID <Y> ^line-3 +)
   --> (remove 4))

(p three-three-minus
   (stage reduce-candidates)
   (junction ^junction-ID <X> ^line-ID-3 <L>)
   (junction ^junction-ID {<Y> <> <X>} ^line-ID-3 <L>)
   (labelling-candidate ^junction-ID <X> ^line-3 -)
  -(labelling-candidate ^junction-ID <Y> ^line-3 -)
   --> (remove 4))

(p three-three-in
   (stage reduce-candidates)
   (junction ^junction-ID <X> ^line-ID-1 <L>)
   (junction ^junction-ID {<Y> <> <X>} ^line-ID-1 <L>)
   (labelling-candidate ^junction-ID <X> ^line-1 in)
  -(labelling-candidate ^junction-ID <Y> ^line-1 out)
   --> (remove 4))

(p three-three-out
   (stage reduce-candidates)
   (junction ^junction-ID <X> ^line-ID-3 <L>)
   (junction ^junction-ID {<Y> <> <X>} ^line-ID-3 <L>)
   (labelling-candidate ^junction-ID <X> ^line-3 out)
  -(labelling-candidate ^junction-ID <Y> ^line-3 in)
   --> (remove 4))
```

```
(p go-to-print-out
   (stage reduce-candidates)
   -->
   (remove 1)
   (make stage print-out))

;;;;;;;;;;;;;;
; Print Out ;
;;;;;;;;;;;;;

(p print-out
   (stage print-out)
   -->
   (remove 1)
   (halt))

;;;;;;;;;;;;;;
; Run        ;
;;;;;;;;;;;;;;

(make start)
(watch 2)
(run)
(ppwm)
```

Bibliography

[Acharya and Tambe, 1989] A. Acharya and M. Tambe, "Production Systems on Message Passing Computers: Simulation Results and Analysis," *International Conference on Parallel Processing (ICPP-89)*, pp. 246-254, 1989.

[Adler *et al.*, 1988] M. R. Adler, A. B. Davis, R. Weihmayer and R. W. Worrest, "Conflict Resolution Strategies in Non-Hierarchical Distributed Agents," *International Workshop on Distributed Artificial Intelligence (DAIW-88)*, USC Computer Research Institute, CRI-88-41, 1988.

[Bernstein and Goodman, 1981] P. A. Bernstein and N. Goodman, "Concurrency Control in Distributed Database Systems," *ACM Computing Surveys*, Vol. 13, No. 2, pp. 185-221, 1981.

[Bomans and Roose, 1989] L. Bomans and D. Roose, "Benchmarking the iPSC/2 Hypercube Multiprocessor," *Concurrency: Practice and Experience*, Vol. 1, pp. 3-18, 1989.

[Bond and Gasser, 1988] A. Bond and L. Gasser, *Readings in Distributed Artificial Intelligence*, Morgan Kaufman, 1988.

[Bowen *et al.*, 1989] T. F. Bowen, F. S. Dworack, C. H. Chow, N. Griffeth, G. E. Herman and Y-J. Lin, "The Feature Interaction Problem in Telecommunications Systems," *International Conference on Software Engineering for Telecommunication Switching Systems*, pp. 59-62, 1989.

[Bridgeland and Huhns, 1990] D. M. Bridgeland and M. N. Huhns, "Distributed Truth Maintenance," *National Conference on Artificial Intelligence (AAAI-90)*, pp. 72-77, 1990.

[Brownston *et al.*, 1985] L. Brownston, R. Farrell, E. Kant and N. Martin, *Programming Expert System in OPS5: An Introduction to Rule Based Programming*, Addison-Wesley, 1985.

[Clayton, 1987] B. D. Clayton, *ART Programming Tutorial*, Inference Corp., 1987.

[Corkill, 1982] D. D. Corkill, *A Framework for Organizational Self-Design in Distributed Problem Solving Networks*, PhD Dissertation, COINS-TR-82-33, University of Massachusetts, 1982.

[Cruise *et al*, 1987] A. Cruise, R. Ennis, A. Finkel, J. Hellerstein, D. Klein, D. Loeb, M. Masullo, K. Milliken, H. Van Woerkom and N. Waite, "YES/L1: Integrating Rule-based, Procedural, and Real-time Programming for Industrial Applications," *IEEE Conference on Artificial Intelligence for Applications (CAIA-87)*, pp. 134-139, 1987.

[Dally, 1986] W. J. Dally, "Directions in Concurrent Computing," *International Conference on Computer Design*, pp. 102-106, 1986.

[Davis and Smith, 1983] R. Davis and R. G. Smith, "Negotiation as a Metaphor for Distributed Problem Solving," *Artificial Intelligence*, Vol. 20, pp. 63-109, 1983.

[Durfee *et al.*, 1987] E. H. Durfee, V. R. Lesser, and D.D., Corkill, "Coherent Cooperation among Communicating Problem Solvers," *IEEE Transactions on Computers*, Vol. 36, pp. 1275-1291, 1987.

[Durfee and Lesser, 1987] E. H. Durfee and V. R. Lesser, "Using Partial Global Plans to Coordinated Distributed Problem Solvers," *International Joint Conference on Artificial Intelligence (IJCAI-87)*, pp. 875-883, 1987.

[Ensor and Gabbe, 1985] J. R. Ensor and J. D. Gabbe, "Transactional Blackboard," *International Joint Conference on Artificial Intelligence (IJCAI-85)*, pp. 340-344, 1985.

[Fennell and Lesser, 1981] R. D. Fennell and V. R. Lesser, "Parallelism in Artificial Intelligence Problem Solving: A Case Study of Hearsay-II," *IEEE Transactions on Computers*, Vol. 26, No. 2, pp. 98-111, 1977.

[Forgy, 1981] C. L. Forgy, *OPS5 User's Manual*, CS-81-135, Carnegie-Mellon University, 1981.

[Forgy, 1982] C. L. Forgy, "RETE: A Fast Algorithm for the Many Pattern / Many Object Pattern Match Problem," *Artificial Intelligence*, Vol. 19, pp. 17-37, 1982.

[Forgy *et al*, 1984] C. L. Forgy, A. Gupta, A. Newell and R. Wedig, "Initial Assessment of Architectures for Production Systems," *National Conference on Artificial Intelligence (AAAI-84)*, pp. 116-120, 1984.

[Forgy, 1985] C. L. Forgy, *OPS83 User's Manual and Report*, Production Systems Technologies, Inc., 1985.

[Fox, 1981] M. Fox, "An Organizational View of Distributed Systems," *IEEE Transactions on Systems, Man, and Cybernetics*, Vol. 11, January, 1981.

[Fukuhara *et al*, 1990] Y. Fukuhara, K. Suzuki, M. Kiyama, H. Okamoto, "Educational Assistant Expert System (CAIRN)," *NTT R&D*, Vol. 39, No. 3, pp. 421-428, 1990.

[Gasser, 1986] L. Gasser, "The Integration of Computing and Routine Work," *ACM Transactions on Office Information Systems*, Vol. 4, No. 3, pp., 205-225, July, 1986.

[Gasser *et al.*, 1987] L. Gasser, C. Braganza and N. Herman, "MACE: A Flexible Testbed for Distributed AI Research," *Distributed Artificial Intelligence*, pp. 119-152, Morgan Kaufmann, 1987.

[Gasser *et al.*, 1989] L. Gasser, N. Rouquette, R. Hill and J. Lieb, "Representing and Using Organizational Knowledge in DAI Systems," Les Gasser and M. N. Huhns, Eds., *Distributed Artificial Intelligence Volume II*, Morgan Kaufman, pp. 55-78, 1989.

[Gasser, 1991] L. Gasser, "Social Conceptions of Knowledge and Action," *Artificial Intelligence*, pp. 107-138, January, 1991.

[Gasser and Ishida, 1991] L. Gasser and T. Ishida, "A Dynamic Organizational Architecture for Adaptive Problem Solving," *National Conference on Artificial Intelligence (AAAI-91)*, pp. 185-190, 1991.

[Georgeff, 1982] M. P. Georgeff, "Procedural Control in Production Systems," *Artificial Intelligence*, Vol. 18, pp. 175-201, 1982.

[Georgeff, 1983] M. P. Georgeff, "Communication and Interaction in Multi-Agent Planning," *National Conference on Artificial Intelligence (AAAI-83)*, pp. 125-129, 1983.

[Georgeff and Lansky, 1986] M. P. Georgeff and A. L. Lansky, "A System for Reasoning in Dynamic Domains: Fault Diagnosis on the Space Shuttle," *SRI International Technical Note 375*, 1986.

[Georgeff and Ingrand, 1990] M. P. Georgeff and F. F. Ingrand, "Real-Time Reasoning: The Monitoring and Control of Spacecraft Systems," *IEEE Conference on Artificial Intelligence for Applications (CAIA-90)*, pp. 198-204, 1990.

[Gupta *et al.*, 1986] A. Gupta, C. L. Forgy, A. Newell and R. Wedig, "Parallel Algorithms and Architectures for Rule-Based Systems," *International Symposium on Computer Architecture*, pp. 28-37, 1986.

[Gupta *et al.*, 1988] A. Gupta, C. L. Forgy, D. Kalp, A. Newell and M. Tambe, "Parallel OPS5 on the Encore Multimax," *International Conference on Parallel Processing (ICPP-88)*, pp. 271-280, 1988.

[Gupta, 1987] A. Gupta, *Parallelism in Production Systems*, Morgan Kaufmann, 1987.

[Griffeth, 1991] N. Griffeth, "The Negotiating Agent Model for Establishing and Modifying Communications," *TINA-91*, 1991.

[Harvey *et al.*, 1991] W. Harvey, D. Kalp, M. Tambe, D. Mckeown and A. Newell, "The effectiveness of Task-Level Parallelism for Production Systems," *Special Issue on the Parallel Execution of Rule Systems, Journal of Parallel and Distributed Computing*, Vol. 13, pp. 395-411, 1991.

[Hayes-Roth *et al*, 1983] F. Hayes-Roth, D. A. Waterman and D. B. Lenat, *Building Expert Systems*, Addison-Wesley, 1983.

[Hayes-Roth, 1985] B. Hayes-Roth, "A Blackboard Architecture for Control," *Artificial Intelligence*, Vol. 26, pp. 251-321, 1985.

[Hayes-Roth, 1987] B. Hayes-Roth, "A Multi-Processor Interrupt-Driven Architecture for Adaptive Intelligent Systems," *Stanford University Technical Report*, KSL 87-31, 1987.

[Hayes-Roth *et al.*, 1989] B. Hayes-Roth, R. Washington, R. Hewett, M. Hewett and A. Seiver, "Intelligent Monitoring and Control," *International Joint Conference on Artificial Intelligence (IJCAI-89)*, pp. 243-249, 1989.

[Hewitt, 1977] C. Hewitt, "Viewing Control Structures as Patterns of Passing Messages," *Artificial Intelligence*, Vol. 8, No. 3, pp. 323-364, 1977.

[Hewitt, 1991] C. Hewitt, "Open Systems Semantics for Distributed Artificial Intelligence" *Artificial Intelligence*, pp. 79-106, January, 1991.

[Hoare, 1987] C. A. R. Hoare, "Communicating Sequential Processes," *CACM*, Vol. 21, No. 8, pp. 666-677, 1978.

[Hogg and Huberman, 1990] T. Hogg and B. A. Huberman, "Controlling Chaos in Distributed Systems," *Xerox Palo Alto Research Center Technical Report*, SSL-90-52, 1990.

[Hsu *et al.*, 1987] C. Hsu, S. Wu and J. Wu, "A Distributed approach for Inferring Production Systems," *International Joint Conference on Artificial Intelligence (IJCAI-87)*, pp. 62-67, 1987.

[Ishida and Stolfo, 1984] T. Ishida and S. J. Stolfo, "Simultaneous Firing of Production Rules on Tree Structured Machines," *Columbia University Technical Report*, CUCS-109-84, 1984.

[Ishida and Stolfo, 1985] T. Ishida and S. J. Stolfo, "Towards Parallel Execution of Rules in Production System Programs," *International Conference on Parallel Processing (ICPP-85)*, pp. 568-575, 1985.

[Ishida, 1989] T. Ishida, "CoCo: A Multi-Agent System for Concurrent and Cooperative Operation Tasks," *International Workshop on Distributed Artificial Intelligence (DAIW-89)*, pp. 197-213, 1989.

[Ishida, 1990] T. Ishida, "Methods and Effectiveness of Parallel Rule Firing," *IEEE Conference on Artificial Intelligence for Applications (CAIA-90)*, pp. 116-122, 1990.

[Ishida et al., 1990] T. Ishida, M. Yokoo and L. Gasser, "An Organizational Approach to Adaptive Production Systems," *National Conference on Artificial Intelligence (AAAI 90)*, pp. 52 58, 1990.

[Ishida, 1991] T. Ishida, "Parallel Rule Firing in Production Systems," *IEEE Transactions on Knowledge and Data Engineering*, Vol. 3, No. 1, pp. 11-17, 1991.

[Ishida et al., 1991] T. Ishida, Y. Sasaki and Y. Fukuhara, "Use of Procedural Programming Language for Controlling Production Systems," *IEEE Conference on Artificial Intelligence Applications (CAIA-91)*, pp. 71-75, 1991.

[Ishida, 1992] T. Ishida, "A Transaction Model for Multiagent Production Systems," *IEEE Conference on Artificial Intelligence for Applications (CAIA-92)*, pp. 288-294, 1992.

[Ishida, 1993] T. Ishida, "Towards Organizational Problem Solving," *IEEE Conference on Robotics and Automation (R&A-93)*, 1993.

[Ishida, 1994] T. Ishida, "An Optimization Algorithm for Production Systems," *IEEE Transactions on Knowledge and Data Engineering*, 1994.

[Ishida et al., 1994] T. Ishida, Y. Sasaki, K. Nakata and Y. Fukuhara, "An Meta-Level Control Architecture for Production Systems," *IEEE Transactions on Knowledge and Data Engineering*, 1994.

[Ishikawa et al., 1987] Y. Ishikawa, H. Nakanishi and Y. Nakamura, "An Expert System for Optimizing Logic Circuits," In *National Convention of Information Processing Society of Japan (in Japanese)*, pp. 1391-1392, 1987.

[Jarke and Koch, 1984] M. Jarke and J. Koch, "Query Optimization in Database Systems," *Computing Surveys*, Vol. 16, No. 2, pp. 111-152, 1984.

[JPDC, 1991] *Special Issue on the Parallel Execution of Rule Systems, Journal of Parallel and Distributed Computing*, Vol. 13, No. 4, 1991.

[Kuo *et al.*, 1991] C. Kuo, D. P. Miranker and J. C. Browne, "On the Performance of the CREL System," *Special Issue on the Parallel Execution of Rule Systems, Journal of Parallel and Distributed Computing*, Vol. 13, pp. 424-441, 1991.

[Kuo and Moldovan, 1991] S. Kuo and D. Moldovan, "Implementation of Multiple Rule Firing Production Systems on Hypercube," *Special Issue on the Parallel Execution of Rule Systems, Journal of Parallel and Distributed Computing*, Vol. 13, pp. 383-394, 1991.

[Kuo and Moldovan, 1992] S. Kuo and D. Moldovan, "The State of the Art in Parallel Production Systems," *Journal of Parallel and Distributed Computing*, Vol. 15, pp. 1-26, 1992.

[Laffey *et al.*, 1988] T. J. Laffey, P. A. Cox, J. L. Schmidt, S. M. Kao, and J. Y. Read, "Real-Time Knowledge-Based Systems," *AI Magazine*, Vol. 9, No. 1, pp. 27-45, 1988.

[Lesser *et al.*, 1988] V. R. Lesser, J. Pavlin and E. H. Durfee, "Approximate Processing in Real Time Problem Solving," *AI Magazine*, Vol. 9, No. 1, pp. 49-61, 1988.

[Laird and Newell, 1984] J. E. Laird and A. Newell, "A Universal Weak Method," *Carnegie-Mellon University Technical Report*, CS-83-141, 1983.

[Lerner and Cheng, 1983] M. Lerner and J. Cheng, "The Manhattan Mapper Expert Production System," *Columbia University Technical Report*, Department of Computer Science, 1983.

[Lesser and Corkill, 1983] V. R. Lesser and D. D. Corkill, "The Distributed Vehicle Monitoring Testbed: A Tool for Investigating Distributed Problem Solving Networks," *AI Magazine*, pp. 15-33, 1983.

[Malone, 1987] T. W. Malone, "Modeling Coordination in Organizations and Markets," *Management Science*, Vol. 33, No. 10, pp. 1317-1332, 1987.

[Mason and Johnson, 1989] C. L. Mason and R. R. Johnson, "DATMS: A Framework for Distributed Assumption Based Reasoning," Les Gasser and M. N. Huhns, Eds., *Distributed Artificial Intelligence Vol.II*, pp. 293 - 318, Morgan Kaufmann, 1989.

[McDermott and Forgy, 1978] J. McDermott, and C. L. Forgy, "Production System Conflict Resolution Strategies," D. A. Waterman and F. Hayes-Roth, Eds., *Pattern Directed Inference Systems*, Academic Press, 1978.

[Miranker, 1987] D. P. Miranker, "TREAT: A Better Match Algorithm for AI Production Systems," *National Conference on Artificial Intelligence (AAAI-87)*, pp. 42-47, 1987.

[Miranker et al., 1990] D. P. Miranker, B. J. Lofaso, G. Farmer, A. Chandra and D. Brant, "On a TREAT Based Production System Compiler," *International Conference on Expert Systems*, Avignon, France, 1990.

[Miranker, 1990a] D. P. Miranker, "An Algorithmic Basis for Integrating Production Systems and Large Databases," *6th IEEE Data Engineering Conference*, 1990.

[Miranker, 1990b] D. P. Miranker, *TREAT: A New and Efficient Match Algorithm for AI Production Systems*, Morgan Kaufmann, 1990.

[Moldovan, 1986] D. I. Moldovan, "A Model for Parallel Processing of Production Systems," *IEEE International Conference on Systems, Man, and Cybernetics*, pp. 568-573, 1986.

[Morihara et al., 1987] I. Morihara, T. Ishida and H. Furuya, "Rule-Based Flexible Control of Tutoring Process in Scene-Oriented CAI Systems," *IEEE Conference on Artificial Intelligence for Applications (CAIA-87)*, pp. 207-212, 1987.

[Nilsson, 1980] N. J. Nilsson, *Principles of Artificial Intelligence*, Tioga, Palo Alto, Calif., 1980.

[Oshisanwo and Dasiewicz, 1987] A.O. Oshisanwo and P.P. Dasiewicz, "A Parallel Model and Architecture for Production Systems," *International Conference on Parallel Processing (ICPP-87)*, pp. 147-153, 1987.

[Raschid et al., 1988] L. Raschid, T. Sellis and C-C. Lin, "Exploiting Concurrency in a DBMS Implementation for Production Systems," *International Symposium on Databases in Parallel and Distributed Systems*, 1988.

[Scales, 1986] D. J. Scales, *Efficient Matching Algorithms for the SOAR/OPS5 Production System*, Stanford University Technical Report, STAN-CS-86-1124, 1986.

[Schor et al., 1986] M. I. Schor, T. P. Daly, H. S. Lee and B. R. Tibbitts, "Advances in RETE Pattern Matching," *National Conference on Artificial Intelligence (AAAI-86)*, pp. 226-232, 1986.

[Sellis and Lin, 1990] T. Sellis and C-C. Lin, "Performance of DBMS Implementation of Production Systems," *IEEE Conference on Tools for Artificial Intelligence (TAI-90)*, pp. 393-399, 1990.

[Schmolze and Goel, 1990] J. G. Schmolze and S. Goel, "A Parallel Asynchronous Distributed Production System," *National Conference on Artificial Intelligence (AAAI-90)*, pp. 65-71, 1990.

[Schmolze, 1991] J. G. Schmolze, "Guaranteeing Serializable Results in Synchronous Parallel Production Systems," *Special Issue on the Parallel Execution of Rule Systems, Journal of Parallel and Distributed Computing,* Vol. 13, pp. 348-365, 1991.

[Smith and Genesereth, 1985] D. E. Smith and M. R. Genesereth, "Ordering Conjunctive Queries," *Artificial Intelligence*, Vol. 26, pp. 171-215, 1985.

[Soloway et al., 1987] E. Soloway, J. Bachant and K. Jensen, "Assessing the Maintainability of XCON-in-RIME: Coping with the Problem of a VERY Large Rule-base," *National Conference on Artificial Intelligence (AAAI-87)*, pp. 824-829, 1987.

[Soo and Schor, 1990] H. Soo and M. I. Schor, "Dynamic Argumentation of Generalized Rete Networks for Demand-Driven Matching and Rule Updating," *IEEE Conference on Artificial Intelligence for Applications (CAIA-90)*, pp. 123-129, 1990.

[Stolfo, 1979] S. J. Stolfo, *Automatic Discovery of Heuristics for Nondeterministic Programs from Sample Execution Traces*, PhD Thesis, Courant Computer Sciences, Rept. No. 18, Courant Inst., New York University, 1979.

[Stolfo and Shaw, 1982] S. J. Stolfo and D. E. Shaw, "DADO: A Tree Structured Machine Architecture for Production Systems," *National Conference on Artificial Intelligence (AAAI-82)*, 1982.

[Stolfo, 1984] S. J. Stolfo, "Five Parallel Algorithms for Production System Execution on the DADO Machine," *National Conference on Artificial Intelligence (AAAI-84)*, pp.300-307, 1984.

[Stolfo et al., 1991] S. J. Stolfo, O. Wolfson, P. K. Chan, H. M. Dewan, L. Woodbury, J. S. Glazier and D. A. Ohsie, "PARULEL: Parallel Rule Processing Using Meta-rules for Reduction," *Special Issue on the Parallel Execution of Rule Systems, Journal of Parallel and Distributed Computing,* Vol. 13, pp. 366-382. 1991.

[Stuart, 1985] C. J. Stuart, "An Implementation of a Multiagent Plan Synchronizer," *International Joint Conference on Artificial Intelligence (IJCAI-85)*, pp. 1031-1033, 1985.

[von Martial, 1989] F. von Martial, "Multiagent Plan Relationships," *International Workshop on Distributed Artificial Intelligence (DAIW-89)*, pp. 59-72, 1989.

[Warren, 1981] D. H. D. Warren, "Efficient Processing of Interactive Relational Database Queries Expressed in Logic," *International Conference on Very Large Databases (VLDB-81)*, pp. 272-281, 1981.

[Winston, 1977] P. H. Winston, *Artificial Intelligence*, Addison-Wesley, 1977.

[Yokoo *et al.*, 1992] M. Yokoo, E. H. Durfee, T. Ishida and K. Kuwabara, "Distributed Constraint Satisfaction for Formalizing Distributed Problem Solving," *International Conference on Distributed Computing Systems (ICDCS-92)*, pp. 614-621, 1992.

[Zisman, 1980] M. D. Zisman, "Using Production Systems for Modeling Asynchronous Concurrent Processes," D. A. Waterman and F. Hayes-Roth, Eds., *Pattern-Directed Inference Systems*, Academic Press, New York, 1978.

Index

Springer-Verlag
and the Environment

We at Springer-Verlag firmly believe that an international science publisher has a special obligation to the environment, and our corporate policies consistently reflect this conviction.

We also expect our business partners – paper mills, printers, packaging manufacturers, etc. – to commit themselves to using environmentally friendly materials and production processes.

The paper in this book is made from low- or no-chlorine pulp and is acid free, in conformance with international standards for paper permanency.

Lecture Notes in Artificial Intelligence (LNAI)

Lecture Notes in Computer Science